Seeing Green

Seeing Green

*The Use and Abuse of
American Environmental Images*

FINIS DUNAWAY

The University of Chicago Press   Chicago and London

FINIS DUNAWAY is associate professor of history at Trent
University, where he teaches courses in US history, visual culture,
and environmental studies.

The University of Chicago Press, Chicago 60637
The University of Chicago Press, Ltd., London
© 2015 by The University of Chicago
All rights reserved. Published 2015.
Printed in the United States of America

24 23 22 21 20 19 18 17 16 15   1 2 3 4 5

ISBN-13: 978-0-226-16990-3 (cloth)
ISBN-13: 978-0-226-16993-4 (e-book)
DOI: 10.7208/chicago/9780226169934.001.0001

Library of Congress Cataloging-in-Publication Data

Dunaway, Finis, author.
  Seeing green : the use and abuse of American environmental
images / Finis Dunaway.
     pages ; cm
  Includes bibliographical references and index
  ISBN 978-0-226-16990-3 (cloth : alk. paper) — ISBN 0-226-16990-1
(cloth : alk. paper) — ISBN 978-0-226-16993-4 (e-book)   1. Environmentalism—United States.   2. Visual communication—United
States.   3. Environmentalism in mass media.   4. Environmentalism
in art.   5. Disasters in art.   I. Title.
  P96.E572U63 2015
  363.701′4—dc23

                                                  2014029025

Publication of this book has been aided by a grant from the Neil
Harris Endowment Fund, which honors the innovative scholarship
of Neil Harris, the Preston and Sterling Morton Professor Emeritus of
History at the University of Chicago.The Fund is supported by contributions from the students, colleagues, and friends of Neil Harris.

♾ This paper meets the requirements of ANSI/NISO Z39.48-1992
(Permanence of Paper).

For Max and Zoe

Contents

CONTENTS

Introduction

Certain images stand out as icons of American environmentalism: a 1971 public service announcement featuring the "Crying Indian," who sheds a tear in response to litter and pollution; the cooling towers of Three Mile Island, site of a notorious nuclear power accident in 1979; the sorrowful spectacle of oil-soaked otters and birds following the 1989 *Exxon Valdez* spill; and, more recently, Al Gore delivering his global-warming slide show in *An Inconvenient Truth*. These images, and others like them, have helped make environmental consciousness central to American public culture.

If you look through most histories of environmentalism, though, you will find few of these images. Standard accounts of the movement emphasize the growth of local and national organizations, the contributions of key thinkers and activists, or the impact of environmentalism on public policy. Despite the important insights yielded by these approaches, such histories have failed to consider the crucial role images have played in the making of popular environmentalism. While traditional histories focus on political struggles, legislative reforms, and scientific writings, this book moves beyond conventional sources to place media images at the center of its analysis. Rather than presenting pictures as mere illustrations, as passive mirrors that simply reflect historical change, I instead consider images as active rhetorical agents. Media images do not simply illustrate environmental politics, but also shape the bounds of public debate by naturalizing particular meanings of environmentalism. As they draw a broader public of media consumers into popular environmentalism, images act as both revelations and veils, creating tensions between what they visualize and what they hide, which ideas they endorse and which they deny.[1]

Seeing Green shows how popular environmentalism has been entwined with mass-media spectacles of crisis. This fusion of politics and spectacle has encouraged Americans to see themselves as part of a larger ecological fabric, and to support personal and political change to protect the environment. Yet, even as media images have made the environmental crisis visible to a mass public, they often have masked systemic causes and ignored structural inequalities. Deflecting attention from corporate and government responsibility, popular images have instead emphasized the idea that individual Americans are personally culpable for pollution and other environmental problems. The visual media have thus offered environmentalists a double-edged sword: Images have helped them popularize their cause, but have also distorted their ideas by portraying their movement as a moralistic crusade to absolve the nation of its guilt. Ultimately, this dual focus on spectacles of crisis and individual consumer choices has hidden underlying causes and structural solutions behind a veil of inattention.

Beginning with radioactive fallout and pesticides during the 1960s and ending with global warming today, this book looks at a wide array of media images—including pictures in popular magazines, television news, advertisements, cartoons, films, and political posters—to explain how dominant ideas of environmentalism became naturalized through repetition. Rather than focus on one genre of representation, such as photojournalism or Hollywood film, I decided to take a broader view and consider the cross-fertilization of ideas and motifs across a variety of mainstream media sources. This approach not only registers the intertextual experience of audiences who encounter visual images in diverse forms and contexts, but also reveals the overarching themes and tropes that have shaped the dominant meanings of popular environmentalism.

Seeing Green emphasizes three broad themes often missing from other histories: the emotions and public life, the shifting meanings of environmental citizenship, and the limits of media representation. First, I explain how the public life of environmentalism has depended upon the power of media imagery to evoke audience emotion and give visible form to fear, guilt, hope, and other environmental feelings. Emotions are not peripheral to politics and public life, but rather play an active role in galvanizing environmental concern. Although it is common to view reason and emotion as diametrical opposites, popular images have fostered politically-charged, scientifically-informed feelings about the environmental crisis. By exploring the fusion of fact and feeling in environmental icons, this book questions the supposed separation between cognition and emotion and shows how scientific knowledge has often been conveyed through visual spectacle.[2]

In tracing the emotional history of environmentalism, I identify a recurring pattern in popular imagery: a focus on children as emotional emblems of the future. Children have long played an important symbolic role in other reform efforts—from Progressive Era campaigns against child labor to civil rights protests in the 1960s—yet their frequent presence in environmental imagery requires its own explanation. Within the context of popular environmentalism, children's bodies provide a way to visualize the largely invisible threats of radiation, toxicity, and other environmental dangers.

The vulnerable child—usually and not coincidentally a white child—became a key visual motif to project a sense of universal vulnerability. This equation of whiteness with universal danger has made environmental problems appear to transcend race and class divisions. Although the poor and racial minorities have often been exposed to higher levels of pollutants and toxicity, media images have repeatedly imagined the citizenry as being equally vulnerable to environmental risk. By depicting white bodies as signs of universal vulnerability, this imagery has mobilized environmental concern but has also masked the ways in which structural inequities produce environmental injustice.

The popular discourse of universal vulnerability relates to my second focus on the meanings of environmental citizenship. I use this term to denote the ecological rights and responsibilities of citizens: from state policies that promise to protect people from toxicity and other environmental risks to individuals engaging in ecologically responsible actions in daily life. I argue that the visual media function as an important technology of environmental citizenship, and I ask how various images have enlarged, restricted, or otherwise defined the scope of ecological rights and responsibilities in modern America.[3]

Popular imagery of universal vulnerability has often been paralleled by the notion of universal responsibility: the idea that all Americans are equally to blame for causing the environmental crisis. Rather than making demands upon the state to ensure citizen rights to a clean, safe environment, this dimension of environmental citizenship focuses on the private sphere of home and consumption and frames personal actions—including recycling, energy conservation, and green consumerism—as essential to saving the planet. This emphasis on individual action often obscures the role of corporations and governments in making the production decisions that result in large-scale environmental degradation. As we will see, appeals to individual responsibility emerged in tandem with the rise of popular environmentalism, and began to circulate widely during the period surrounding the first Earth Day in 1970. In recent decades, with the advance of neoliberalism—meaning the revival of classic, eighteenth-century liberalism's focus on free markets and deregulation—this model of environmental citizenship based on green consumerism has become increasingly triumphant. American environmental citizenship—especially the lopsided faith in personal action and green consumerism—has been powerfully transacted through visual images that imagine the political world in an individualist frame, that mobilize feelings of fear and guilt to instill a sense of personal responsibility for the environment.[4]

Popular framings of environmentalism that link emotions to citizenship lead directly to this book's third major theme: the limits of media representation. Even as visual images provide important resources for democratic politics, they also work to constrict the imagination of the political world. Even as they expand conceptions of citizenship to include environmental rights and responsibilities, they often narrow the scope of action to emphasize immediate reforms or consumer decisions and thus foreclose on other possibilities for change. Finally, even as they publicize moments of crisis, media images often detach dramatic episodes from the broader contexts and timescales of ecological danger. Looking closely at how images negotiate three crucial issues—environmental time, power relations, and possible solutions—reveals how the media have both defined and delimited the scope of popular environmentalism.[5]

Unlike the apocalyptic scenarios associated with thermonuclear devastation, the catastrophic visions of environmentalism often emphasize a longer time frame of fear, a mode of risk based upon the gradual accumulation of hazardous agents in the human body and in the soil, water, and atmosphere. Rather than the sudden, immediate destruction unleashed by the bomb, the problems of the environmental crisis—such as air and water pollution, pesticide buildup in the food chain, and greenhouse gas emissions—often suggest a long-term, slowly escalating sense of danger. The literary critic Rob Nixon terms this incremental form of ecological calamity "slow violence": "a violence that occurs gradually and out of sight, a violence of delayed destruction that is dispersed across time and space, an attritional violence that is typically not viewed as violence at all." Indeed, the material realities of slow violence pose a representational challenge for environmentalists as they engage with the spectacle-driven dictates of the mass media, as they seek to move beyond the temporality of immediate catastrophe to warn of incremental crises in the making.[6]

Seeing Green identifies tensions within the mainstream media's attempts to frame environmental time, and considers how they have proven to be both productive and problematic for environmental politics. On the one hand, visual images have played a crucial role in legitimating the concept of environmental crisis and, at times, have encouraged audiences to see even spectacular moments of catastrophe—such as the 1969 Santa Barbara oil spill—not as isolated or aberrant phenomena, but rather as signs of an all-encompassing, gradually escalating calamity. Sometimes this coverage has also worked to extend the time frame of citizenship, to invite spectators to glimpse beyond the news and elections cycles and grapple with the intergenerational rights of children and future Americans to a clean and sustainable environment. On the other hand, certain media spectacles have concentrated so much attention on the sense of immediate danger—on, for example, the threat of a meltdown at Three Mile Island—that they have obscured the long-term risks associated with other environmental hazards. Moreover, even when the mainstream media warns of long-term, accretive problems, the proposed solutions often short-circuit time by imagining quick, immediate strategies—consumer actions, technical fixes, or piecemeal legislative reforms—to overcome the crisis.

The limiting power of visual media has been particularly acute for environmental thinkers and activists who reject the rhetoric of universal vulnerability and responsibility to emphasize instead the power relations that structure environmental problems. Media framings that imagine everyone as being equally susceptible to ecological danger have created a unifying vision of environmental citizenship that overlooks environmental injustice. Environmentalism has often been critiqued for its narrow social agenda and for failing to reach beyond its primarily affluent white constituency. *Seeing Green* seeks to understand the powerful role of the visual media in helping produce this exclusionary vision of popular environmentalism. Rather than being straightforward reflections of environmental values, iconic images perform crucial ideological work and often marginalize radical, system-challenging perspectives. At various points in this book, I place mainstream images in dialogue with radical ideas and social movements to reveal the alternative visions that have been ignored or cast aside by our image-driven public culture.[7]

Just as many environmentalists have challenged the discourse of individual responsibility and have offered structural explanations for the causes of the environmental crisis, they have also tried to fashion alternative visions of the future that emphasize systemic solutions to long-term problems. From debates over energy issues in the 1970s to struggles against toxicity and pesticides in the 1980s and beyond, they have repeatedly proposed large-scale changes that have tended to be mocked or marginalized by the mainstream media. This process of filtering out—or altogether ignoring—far-reaching proposals for environmental change demonstrates a significant limit of media imagery. Ultimately, the iconic images of American environmentalism have impeded efforts to realize—or even imagine—sustainable visions of the future.

For the past five decades, the growth of environmentalism has been entwined with media spectacles of environmental crisis. Yet the goals and ideas of environmental activists have not always corresponded with the conventions of media coverage. From nuclear accidents and oil spills to pesticide scares and toxic threats, depictions of crisis have heightened popular concern for particular manifestations of ecological risk, but have failed to communicate more far-reaching ways to confront larger, slowly escalating problems—including the hazards of industrial agriculture, the proliferating presence of toxins in the environment and in human bodies, and the ongoing, increasing dependence upon nonrenewable sources of energy. The limits of media representation can be discerned through the ways in which dominant media institutions have concentrated public attention on certain images of crisis but have tended to ignore, as the philosopher Slavoj Žižek puts it, "the often catastrophic consequences of the smooth functioning of our economic and political systems."[8]

The history recounted here focuses largely on the producers of popular environmental images—including photographers, advertisers, cartoonists, filmmakers, news broadcasters, and environmental organizations—but also tries to understand how consumers have received and responded to these images. I use a variety of evidence—letters to newspaper and magazine editors, film reviews, and archival sources—to consider how audiences have interpreted and made meaning out of environmental icons. Although I offer broad assessments of audience reception, I give particular attention to the responses of two groups: self-defined environmentalists and conservative commentators. As self-described activists, as leaders of environmental organizations, or as influential writers and thinkers, environmentalists have formed a distinct group of media consumers who have frequently challenged and critiqued mainstream depictions of their cause. Although the media have helped popularize environmental concern, activist critiques reveal the limits of media representation and suggest the difficulty of conveying radical ideas through dominant channels of communication. In certain chapters, I also consider the response of conservative pundits, who have frequently lambasted the media for supposedly duping the public into accepting environmentalist claims about the hazards of nuclear power, pesticides, and other issues. Evoking the familiar dualism between reason and emotion, they have presented themselves as the guardians of scientific fact and dismissed environmentalists as the hucksters of spectacle-driven feeling. These debates between environmental activists, conservative commentators, and other viewers demonstrate that images have played a vital and contested role in the public life of environmentalism.

Rather than offering a comprehensive narrative of environmental images, *Seeing Green* takes a selective approach that tries to balance broad scope with careful attention to specific images and the contexts in which they appeared. Beginning in the 1960s, the first chapter offers a prehistory of environmental icons and considers how images of nuclear fallout and pesticides depicted a long-term, gradually escalating sense of ecological danger. From there, the book focuses on three major environmental moments—periods of intense media coverage in which a cluster of images both advanced and constrained the meanings of popular environmentalism. Part 1 explores visual depictions of environmental crisis during the period surrounding Earth Day 1970, including images of the Santa Barbara oil spill, pictures of people wearing gas masks, the Crying Indian, and other texts that engaged with such issues as universal vulnerability and the question of responsibility. Part 2 considers energy crises during the 1970s and shows how a wide range of visual media—including the Hollywood film *The China Syndrome*, media coverage of Three Mile Island, and popular portrayals of solar energy—galvanized public fear of nuclear power but, for the most part, did not give serious attention to questions of systemic overconsumption or the development of renewable energy sources. Part 3 looks at a series of media events—including the *Exxon Valdez* oil spill, the Alar crisis in 1989, and the 1990 Earth Day celebration—to examine the mainstreaming of green values and the triumph of neoliberal environmental citizenship. The conclusion focuses on the surprising popularity of *An Inconvenient Truth* as a way to grapple with the visual politics of environmentalism in our own time.[9]

In these pivotal moments, media images provoked environmental anxiety but also prescribed limited forms of action. In each case, popular images and emotional politics contributed to the making and unmaking of power in the public realm. Each part of this book is divided into several chapters, and each chapter, in turn, focuses primarily on a particular visual text or image motif that became particularly meaningful during a specific moment in the history of environmentalism. Each chapter tells the story of the production and reception of that visual source—a film, political cartoon, advertisement, or set of photographs or other images—and shows how these pictures connect to larger evolving debates over environmental politics. The chapters and parts build thematically and chronologically upon one another, so that the separate stories coalesce as a narrative of connected moments and images in the making of popular environmentalism. By presenting the history of environmentalism as a complex layering of cultural, political, and visual practices, *Seeing Green* explains how images have popularized the cause but have also left crucial issues outside of the frame.

ONE

Dr. Spock, Daisy Girl, and DDT: A Prehistory of Environmental Icons

Everything about him seems so serious: his stiff posture, his stern expression, his three-piece suit, taut necktie, and collar pin (fig. 1.1). With hands in pockets, his lips tightly pursed, he looks down at the child, who seems completely unaware of his presence. Below the photograph, a brief sentence in bold letters summarizes the scene: "Dr. Spock is worried."[1]

Published as a full-page advertisement in the *New York Times* in 1962, reprinted in seven hundred newspapers and numerous magazines, and then appearing as a poster "in store windows, nurseries, doctors' offices, and even on baby carriages," the Dr. Spock ad became the most important visual text produced in the campaign against nuclear testing. As the nation's leading child expert and the author of *Baby and Child Care*, the best-selling parenting manual ever published, Dr. Benjamin Spock exerted tremendous influence in postwar America. His legendary book began with these words of reassurance: "Trust yourself. . . . What good mothers and fathers instinctively feel like doing for their babies is usually best." Although his manual tried to inspire parental confidence, now Spock evinced concern, indeed outright worry, about the dangers of bomb testing and radioactive fallout. He urged parents to protest nuclear testing by writing to their elected

FIGURE 1.1. "Dr. Spock is worried." SANE advertisement, 1962. Courtesy of SANE Inc.
Records, Swarthmore College Peace Collection.

officials and to contribute to SANE, the National Committee for a Sane Nuclear Policy.[2]

Dr. Spock and other SANE ads challenged the sublime aesthetic of the mushroom cloud, the iconic rendering of the bomb blast that celebrated its technologically generated, awe-inspiring qualities. Often produced by government agencies and circulated by *Life* and other popular magazines, photographs of the Nevada Test Site aestheticized the blast and encouraged spectators not to worry about the dangers of radioactive particles released into the atmosphere. By moving beyond the mushroom cloud, leaving the spectacle of the blast to follow radioactivity as it contaminated the environment and entered people's bodies, SANE rejected the government and mass media's framing of the bomb to focus instead on its ominous afterlives. In particular, the organization sought to picture strontium 90 and other radioactive agents as posing a grave, long-term danger to innocent children. Combining empirical data with emotional concern, SANE used visual images to call for environmental citizenship: to demand that the state guarantee the safety of people's living conditions and the futurity of the nation's environment.

SANE ads would soon be joined by other popular images that depicted the temporality of the environmental crisis—that conveyed the long-term danger of radioactive fallout, pesticides, and other threats to the environment and the human body. This chapter will present three sets of images as together constituting a prehistory of environmental icons: SANE ads, the Daisy Girl and other TV commercials produced for the 1964 Lyndon B. Johnson presidential campaign, and pesticide imagery that followed publication of Rachel Carson's *Silent Spring* (1962) and culminated with the 1972 federal ban on DDT. Although these images are not usually viewed in relation to one another, they reveal similar representational strategies and provide new insight into the emergence of modern environmentalism. In these images, the materiality of environmental risk merged with the emotionality of environmental politics to popularize a way of seeing that placed human bodies and nonhuman nature in a shared, interlinked realm of escalating danger. We can think of this as seeing through an ecological lens.

The ecological lens differed from popular landscape photography, which tended to depict wild nature as a pristine realm, untouched and unspoiled by the presence of people. In their campaigns to save wild places, the Sierra Club and other leading wilderness groups used the photography of Ansel Adams and other artists to present nature as a pure, sacred space. Coffee table books, posters, and calendars all celebrated the wilderness as a landscape apart from human society. In contrast, the ecological lens enveloped both humans and nature in a common geography of long-term, incremental danger. SANE ads, LBJ commercials, and pesticide imagery warned of the risks to people and the nonhuman world in ecological systems increasingly burdened by pollution, toxicity, and other hazards. These images helped bring the ecological lens into the mainstream of American public culture.[3]

The popular media sanctioning of environmental values depended upon an important shift in the emotional politics of Cold War America. In the post–World War II period, government officials and scientific experts sought to discredit fear and anxiety as illegitimate emotions, especially in relation to technology, the environment, and human health. They denied the vulnerability of permeable, ecological bodies and urged the public not to be afraid of bomb testing and the proliferation of pesticides and other toxic chemicals. This discourse came under attack, though, as antinuclear activists and ecological critics like Rachel Carson questioned the much-vaunted rationality of the experts and used scientific evidence to justify feelings of fear toward fallout and pesticides. Visual images played an active role in this struggle and ultimately helped to legitimate a new emotional style in American public culture. As fear and anxiety toward the environmental future became normalized, environmental citizenship rights could be more easily imagined, articulated, and realized. This change in emotional styles thus contributed to the emergence of popular environmentalism.

Seeing through the ecological lens abetted a transformation in popular thought, marking a shift from the idea of nature as a realm separate from human society to the notion of environment as an interconnected system that all beings—human and nonhuman—shared. Dr. Spock, the Daisy Girl, and DDT imagery all configured humans as part of nature by showing vulnerable bodies at risk and calling for an emotional response in viewing audiences. These texts fused fact and feeling to frame environmental danger as a slowly escalating crisis, and helped Americans see themselves as part of a living, but also threatened, body politic. In challenging the Cold War emotional style, this new form of popular environmentalism collapsed the boundaries between nature and culture to envision fragile bodies and ecosystems under siege.

By representing fear of fallout, Dr. Spock and other SANE ads sought to reposition iconic photographs of the Nevada Test Site, where the US government detonated approximately one hundred nuclear weapons into the atmosphere during the 1950s and early 1960s. Popular images of the nuclear testing program drew on the sublime tradition to generate a new category of representation: the atomic sublime, a way of seeing that emphasized the overwhelming power and beauty of the mushroom cloud. Rather than picturing vast mountains, towering waterfalls, and other monumental scenes of nature, the atomic sublime naturalized the mushroom cloud to celebrate the spectacle of the blast. As government scientists downplayed the dangers of fallout, the iconic mushroom cloud provided spectators with sublime pictures of US technological power. By encouraging audiences to see beauty but not destruction, to feel inspired but not terrified, to glimpse the mushroom cloud but not contemplate the lingering effects of fallout, the pictures excluded from the frame the threats that strontium 90 and other radioactive elements posed to the environment and human health.[4]

This way of depicting the bomb denied human vulnerability to ecological risk. Indeed, government experts repeatedly claimed that the body was impervious to radioactive materials. One military film, screened for troops before a bomb test in Nevada, shows an animated cartoon of a human body being exposed to radioactive particles. The round objects simply bounce off the skin; unable to penetrate the flesh, they look like balls being thrown against a wall. Merging corporeal fortitude with emotional fortitude, the invincibility of the human body with the rejection of fear, Cold War imagery sought to control public feelings by disavowing environmental danger. Presenting themselves as the purveyors of fact and guardians of rationality, government experts sought to manage public fear by maintaining the boundaries between nature and culture and denying the reality of environmental risk.[5]

This federal dismissal of fallout fear appeared across a diverse array of visual media. In one episode of *The Big Picture*, an Army-produced television show, audiences witnessed the fears of American troops, who worried that their proximity to nuclear tests and exposure to fallout might jeopardize their health. In one scene, supposedly filmed the night before a bomb test, two soldiers speak with a chaplain, who calmly tries to assuage their fears. He first explains how the Army's scientific experts have "taken all the necessary precautions to see that we're perfectly safe here." He then describes the aesthetic pleasure of the bomb blast. "First of all," he says, "one sees a very, very bright light. . . . And then you look up and you see the fireball as it ascends to the heavens. It contains all the rich colors of the rainbow. And then as it rises up into the atmosphere, it . . . assembles into the mushroom. It is a wonderful sight to behold."[6]

In this nationally televised conversation, the chaplain instructed soldiers to put their faith in the experts. Rather than succumbing to fear, they should instead surrender themselves to the intense feelings of awe and wonder summoned by the atomic sublime. By showing the American public how the Army addressed the fears of its troops, *The Big Picture* used TV as an instrument of emotional containment. In these and other examples, government officials and the mass media celebrated the sublime spectacle of the bomb blast but concealed the accretive dangers of radioactivity.

In 1957, SANE announced its founding via a full-page advertisement in the *New York Times*, an ad that looked strikingly different from the Dr. Spock ad that followed five years later (fig. 1.2). The statement began, in large, bold print, with a warning—"We Are Facing A Danger Unlike Any Danger That Has Ever Existed"—and emphasized the ongoing "contamination of air and water and food, and the injury to man himself" caused by nuclear testing. Hoping to build a national movement, but worried that this ad was not attracting enough attention, SANE leaders hired a communication consultant to evaluate its effectiveness. The consultant concluded that the ad was "'too long' and 'too wordy'" and argued "that a photo or graphic symbol would attract the attention of the general public." Indeed, SANE's first foray into the realm of mass communication was completely devoid of images, as lengthy text crowded the page. Even avid readers of the *Times*, the consultant explained, "had 'missed' the ad," not even noticing it was there.[7]

SANE leaders soon followed the consultant's advice. In their initial use of images, they relied upon familiar pictures of the mushroom cloud, recasting the bomb blast as harbinger of doom rather than sublime spectacle. The first illustrated ad began with a warning above the picture: "NUCLEAR BOMBS CAN DESTROY ALL LIFE IN WAR." A second warning appeared below the photograph: "NUCLEAR TESTS ARE ENDANGERING OUR HEALTH RIGHT NOW." Through this fusion of text and image, SANE presented the potential devastation of nuclear war and the actual threat of radioactive fallout as conjoined hazards that both endangered the health of the citizenry.[8]

SANE continued to use mushroom cloud imagery, but some activists urged the group to develop a different representational strategy focusing on childhood vulnerability. Two weeks after the first mushroom cloud ad appeared, one letter writer urged the group to place "ads . . . that starkly present the problem," including, she suggested, "a picture of a child with the caption: THIS CHILD WILL DIE OR BE DEFORMED BY CONTINUING NUCLEAR TESTS: WILL SHE BE YOURS?" This letter writer,

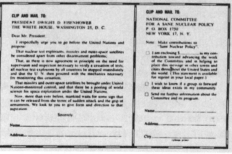

FIGURE 1.2. "We Are Facing a Danger." SANE advertisement, 1957. Courtesy of SANE Inc. Records, Swarthmore College Peace Collection.

along with others who wrote to SANE, believed that visual images, particularly pictures of children, would align emotion with dissent to foster political activism. "Perhaps these . . . suggestions are too macabre," she concluded, "but I feel that public apathy must be met by the most dramatic presentation of this frightening subject." A vibrant democratic culture, along with the future health of Americans and their environment, depended upon, these letter writers believed, the infusion of emotions into politics.[9]

People who recommended that SANE deploy images of children suggested that environmental citizenship could best be imagined through a strategy that depicted the child as the nation's most treasured citizen. These pictures of futurity would ask the presumed audience of adults to worry not so much about the potential risks to their own bodies, but rather to invest concern in the vulnerable bodies of children growing up in an increasingly degraded environment. This representational strategy can be compared with what the cultural critic Lauren Berlant has termed "infantile citizenship." Focusing on the 1980s, Berlant argues that public life and the image of citizenship became overwhelmingly fixated on future Americans, on "pictures . . . circulating in the public sphere" of children, "persons that, paradoxically, cannot yet act as citizens." Yet while infantile citizenship denies adult agency and shifts attention from pressing social problems, SANE's emphasis upon vulnerable children contested the technocratic assumptions of government elites and visualized threats to permeable ecological bodies. By depicting bodies at risk, SANE tried to illustrate the long-term incremental danger of fallout and to challenge the spectacle of the bomb blast through the counterspectacle of innocent children.[10]

In 1962, SANE finally adopted the strategy recommended by letter writers by producing "Dr. Spock is Worried" and other images that visualized the invisible danger of fallout. Hovering over the child, brooding over the dangers of fallout, Spock appears unsure how to protect the child. The text below specifically addresses parents: "If you've been raising a family on Dr. Spock's book, you know that he doesn't get worried easily." "I am worried," Spock explained. "As the tests multiply, so will the damage to children."[11]

This ad bears the hallmarks of the agency that created it: Doyle, Dane, and Bernbach. Known for its distinctive and much-discussed campaigns, like the "Think Small" and "Lemon" ads for the Volkswagen Beetle, Doyle, Dane, and Bernbach departed from the layout style of other firms by using black-and-white photographs, minimalist design, and brief text, usually in the form of short, vertical columns at the bottom of the page. The Dr. Spock ad adopted this same approach, but, unlike other ads by the firm, did not employ humor to make its point.[12]

Many people wrote letters to praise SANE for adopting this strategy which appealed to audience emotions. "The Dr. Spock advertisement is the greatest [SANE] has ever published," one writer announced. "It represents a sure-fire way of reaching people who would otherwise not be touched by even the most logical approach." Many believed that Spock's celebrity status and the respect that parents, especially mothers, accorded him would enable SANE to reach a larger public. "You know," one woman wrote, "the most faithful readers of advertisements are housewives looking for items for the home and family. For that reason, the Dr. Spock ad is by far the most effective ad SANE has ever sponsored." She also applauded the group for using a photograph and brief verbal appeal rather than relying too heavily on text. "The well-documented SANE ads with a couple of 1000 words of copy may be literary masterpieces," she continued, "—but the people who read ads as a rule aren't apt to read them!" Others described Spock as "practically a member of the family." "I was very moved by your statement," one explained. "Mothers have looked to you for advice regarding children for years. I sincerely hope they heed your voice at this time."[13]

Jeanne Bagby, a cofounder of the antinuclear group Women Strike for Peace, also expressed her gratitude for the ad. "I wish to join with thousands of other mothers in heartfelt thanks for your recent public statement," Bagby wrote to Spock. "For months and years, we have implored our doctors, scientists and government officials for the truth about radiation effects, for a more public discussion of the hazards of testing, for protection of our children. Our efforts have been largely dismissed as the rantings of hysterical women." According to Bagby, women protesters were too often ignored or ridiculed on the basis of the misogynistic charge that they were overly emotional and thus unable to cope rationally with the arms race and the debate over fallout. For this reason, she suggested that a man could more easily convey emotional concern and not be dismissed.[14]

Spock's facial expression in the ad seems to validate his ability to perform this emotive role in an acceptable manner. He does not, after all, appear excessively emotional. He offers solemn concern and informed judgment. Even as his anxieties are "written all over his kindly face," as *Time* magazine put it, Spock still maintains his composure, still enacts his masculine authority. In a culture of containment, he still seems contained, still possesses the capacity to impart his expertise.[15]

Other SANE images extended Dr. Spock's warning. An advertisement sponsored by Dentists for SANE showed three youngsters, all sporting open-mouthed smiles and shiny white teeth (fig. 1.3). The text below, though, undercuts the carefree image. "*Your* children's teeth," it warns, "contain Strontium-90." Delivered with scientific certitude and directed once again at parents, this ad referenced the findings of the Baby Teeth Survey conducted by the Committee for Nuclear Information (CNI), a group led by Barry Commoner and other scientists. The sentimental portrait of innocent children unaware of the danger lurking in their bodies provided a visual counterpoint to the information presented in the text. By presenting the children as living in no particular place and by emphasizing that all young Americans are vulnerable to the effects of strontium 90, the ad conveys the idea of universal danger. SANE packaged CNI's information in an explicitly affective form to appeal to parental anxieties about radioactivity.[16]

Every one of the more than two hundred dentists who signed the ad was a man—perhaps not surprising, given that dentistry remained a male-dominated profession. Yet the signers' insistent maleness subtly underscored the gendered politics of SANE. Using their scientific knowledge to prop up their emotional selves, like Spock, all these male dentists expressed their anxiety about the future and their concern about children's ecological bodies. "As dentists, we deplore the buildup of radioactive Strontium-90 in children's teeth and bones," they wrote, in language that upset the dominant mode of detached expertise. "It is a measure of the sickness of our times." Scientific expertise, social responsibility, and smiling children converged in this ad to proclaim that every American, including every dad and every dentist, should demand an end to nuclear testing.[17]

SANE images presented white children as signs of the future and as emblems of universal vulnerability. Their whiteness, of course, went unremarked. Yet the repeated emphasis upon white people, together with the placelessness of the scenes represented—nobody in particular inhabiting nowhere in particular—worked to make the fallout danger seem even more generalized and all-encompassing. No place, the ads asserted—even the lily-white suburbs, those postwar bastions of prosperity and privilege—were immune to the risk of strontium 90 and other radioactive agents. While suburban communities were often depicted as hermetically sealed, blissfully insulated from the problems and perils of public life, these images of white vulnerability overturned dominant ideological assumptions about the human body, the environment, and public culture. Antinuclear groups mobilized scientific knowledge and joined fact with feeling to depict both men and women

FIGURE 1.3. "Your children's teeth contain Strontium-90." SANE advertisement, 1963. Courtesy of SANE Inc. Records, Swarthmore College Peace Collection.

voicing concern about the fate of vulnerable bodies in an atomic age. Their vision of the ecological body blurred the boundaries between nature and culture to portray humans as being situated within threatened ecological systems.[18]

In 1964, the Lyndon B. Johnson presidential campaign appropriated SANE's visual motifs in TV commercials that conveyed fear of fallout and the arms race. The iconic "Daisy Girl" shows a young girl pulling off the petals of a flower, counting each one in ascending order (fig. 1.4). Situated in nature—an icon of innocence, a child who makes a few mistakes as she recites her numbers—she seems, like the flower she clutches, to represent life and futurity. The commercial, also prepared by Doyle, Dane, and Bernbach, again selected a white child to embody universal vulnerability. Meanwhile, the imposing, mechanical-sounding voice of a male narrator begins to count in descending order, from ten to one. The camera moves in closer to the girl's face until images of the mushroom cloud appear on the screen. "Daisy Girl" turns the sublime spectacle of the bomb blast into an image of apocalyptic destruction; it does so by pairing the intimate with the frightening, the private with the public, the innocent with the overwhelming.[19]

Doyle, Dane, and Bernbach produced other commercials for the Johnson campaign, including one that showed a little girl licking an ice cream cone (fig. 1.5). A few seconds into the commercial, an unseen female narrator, presumably a mother figure, begins to speak: "Do you know what people used to do? They used to explode atomic bombs into the air. Now, children should have lots of vitamin A and calcium. But they should not have any strontium 90 or cesium 137. These things come from atomic bombs, and they're radioactive. They can make you die."[20]

After describing the long-term risks of radioactivity, she then explains how the 1963 Limited Test Ban Treaty, which put an end to aboveground nuclear testing, solved this problem and made "the radioactive poisons . . . go away." "But now," she warns, "there is a man who wants to be president of the United States, and he doesn't like this treaty. . . . He wants to go on testing more bombs. His name is Barry Goldwater, and if he's elected, they might start testing all over again." Viewed together, the two commercials—"Daisy Girl" and "Little Girl, Ice Cream Cone"—merged the sudden, immediate horror of nuclear war with the gradual accumulation of fallout in the environment and in children's bodies.

Airing about a year after the signing of the Limited Test Ban Treaty,

FIGURE 1.4. "Daisy Girl." Frame capture from television commercial, 1964. Courtesy of the Lyndon B. Johnson Library and the Democratic National Committee.

FIGURE 1.5. "Little Girl, Ice Cream Cone." Frame capture from television commercial, 1964. Courtesy of the Lyndon B. Johnson Library and the Democratic National Committee.

these commercials visually codified the idea of environmental citizenship and confirmed the importance of continuing the ban. Eight years earlier, when Democratic presidential nominee Adlai Stevenson had argued that the United States should "take the lead in halting further test explosions," his claims were dismissed as being outside the mainstream. In 1964, though, Johnson lauded the treaty for keeping American children protected from radioactive danger. "The deadly products of atomic explosions were poisoning our soil and our food and the milk our children drank and the air we all breathe," he explained. "Radioactive deposits were being formed in increasing quantity in the teeth and bones of young Americans. Radioactive poisons . . . were a growing menace to the health of every unborn child." Johnson's comments reinforced the key claims made by SANE and other groups: that human beings inhabited vulnerable, ecological bodies; that strontium 90 was becoming lodged, in frightening quantities, in the bones and teeth of young children; and that the protection of futurity demanded that nuclear tests no longer be performed above ground.[21]

It is important to note that these two commercials offered different temporal framings of violence. While "Daisy Girl" adhered to notions of spectacular catastrophe and visualized the mushroom cloud as a sign of thermonuclear apocalypse, "Little Girl, Ice Cream Cone" warned of the delayed effects of radioactivity. "Daisy Girl" most likely was understood by audiences at the time as a frightening portrayal of nuclear war—not as a depiction of the slow violence of environmental crisis. Still, Doyle, Dane, and Bernbach's "Little Girl, Ice Cream Cone" reveals that the long time frames of fear—the temporality of environmental crisis—were beginning to seep into American visual culture.

Even as the test ban put an end to the aboveground tests that spewed radioactive agents into the atmosphere, the arms race would continue as the tests descended to subterranean locales. The continued manufacture and testing of bombs would create a series of long-term risks—including uranium mining, weapons production, and underground tests that contaminated workers' bodies and nearby environments—almost all of which remained invisible to the general public. Even though the arms race continued, and even though the production of nuclear weapons created ongoing environmental problems, the sense of urgency about the testing issue soon faded from view. The generalized pictures of universal vulnerability circulated by SANE obscured the specificity of environmental risk by erasing locality and offering viewers a counterspectacle of innocent children in no particular place. While the test ban certainly reduced many children's exposure to strontium 90, it also underwrote a broader pattern of inattention to the disparate spaces where radioactivity continued to enter the environment and people's bodies. The notion of universal victimhood—of every child in every place equally threatened by fallout—masked the ongoing presence of the bomb in the lives of many Americans.[22]

In 1963, over ten million viewers of *CBS Reports*, a highly regarded television news documentary series, encountered a similar image of universal vulnerability: Four children walk through a landscape that resembles the sets of *Leave It to Beaver, Father Knows Best*, and other popular suburban sitcoms. As they stroll past single-family homes and manicured lawns, the children are trailing directly behind a slow-moving truck. The sprayer on the back of the truck releases a thick cloud of chemicals. The children, all of whom are white, continue to amble along, watching the pesticides float across lawns and drift between houses, inhaling toxicity with their every breath (fig. 1.6). As TV audiences glimpsed the children sauntering through this archetypal scene of suburban normalcy, as they stared at the pesticide fog filling

FIGURE 1.6. Children and pesticide truck. Frame capture from *CBS Reports* television documentary "The Silent Spring of Rachel Carson," 1963.

their television screens, they heard the voice of Rachel Carson, who warned, "Can anyone believe it is possible to lay down such a barrage of poisons on the surface of the earth without making it unfit for all life? They should not be called 'insecticides,' but 'biocides.'"[23]

While viewers today would no doubt find these images of children walking through a pesticide cloud shocking and disturbing, similar scenes had been depicted throughout the 1940s and 1950s as normal and acceptable. In 1945, *National Geographic* featured a photograph of a boy smiling and frolicking at Jones Beach (on Long Island, New York), as a truck displaying the sign "D.D.T. / Powerful Insecticide / Harmless to Humans" enshrouds the child with its poisonous spray (fig. 1.7). Three years later, *Life* magazine, in a story that celebrated the aerial spraying of suburban neighborhoods, showed a bikini-clad model clutching a hot dog and drinking a soda as she stares through a fog of DDT. Just as the iconic mushroom cloud sought to deflect fear of fallout and deny ecological risk, these pesticide images rejected the idea that human bodies or ecosystems might be vulnerable to toxicity.[24]

Published just a few months after the Dr. Spock ad appeared, Rachel Carson's *Silent Spring* asked readers to question whether these scenes—fields and forests sprayed with DDT, helicopters showering residential neighborhoods with poison, children walking beside pesticide plumes—should instead be seen as ominous signs of danger. By injecting a long-term sense of fear into popular representations of DDT and other pesticides, Carson warned about escalating threats to futurity. Like SANE, Carson portrayed the home and the private realm as threatened by toxicity, by the assault of deadly chemicals on often unsuspecting citizens. Her vision of environmental citizenship placed SANE's focus on the human body within a broader frame. Carson incorporated the buzzing, blooming world of nature into her sphere of ethical concern to argue that environmental citizenship included the rights of people to live in a healthy environment and to enjoy the scenic beauty of the natural world.

Carson's critique of the pesticide industry provoked a vicious counterattack: a misogynistic effort orchestrated by chemical companies to dis-

FIGURE 1.7. Beachgoers sprayed with DDT at Jones Beach State Park, Long Island, New York, 1945. Copyright Bettmann/Corbis.

miss her claims, often by discrediting her credentials. Male scientists, especially those employed by the chemical industry, openly espoused sexism to charge that as a woman, Carson could not understand the complexity of science. They claimed that she merely responded to the pesticide issue with her emotions rather than with reason. Chemical industry journals published a barrage of negative reviews and scathing editorials; some of their sexist assertions would be trotted out by mainstream reviewers. *Time* magazine, for example, belittled Carson's book as an "emotional and inaccurate outburst" and chastised its author for her "patently unsound" claims and "hysterically overemphatic" prose.[25]

While gendered condemnation shaped the book's reception, what is more striking and surprising is that many of Carson's ideas gained legitimacy within American public culture. Carson negotiated the media landscape and broader culture of images to challenge male proponents of reigning pesticide policy by presenting herself as a reliable, trustworthy source, a countermodel of expertise. The representation of Carson's ideas in various media forms helped reveal the long-term incremental threat that DDT and other pesticides posed to humans and nature.

CBS Reports marked Carson's most important media appearance. Founded in 1959, the news documentary series enjoyed special status on American television screens. The respect accorded to *CBS Reports* would be heightened during times when the show appeared to extricate itself from commercial pressures to inform viewers about controversial issues. The *Silent Spring* episode would become a prime example: Not wanting to be associated with the pesticide story, three sponsors withdrew their commercials. Despite this significant loss of corporate advertising, CBS aired the program and received tremendous praise for not bowing to business demands. "Developments such as this," one critic noted, ". . . only point up the weakness of the sponsor system in American commercial television. The ideal program, from the sponsor's point of view, is a pleasant story about pleasant people with pleasant problems. That's why so much of our programming remains a vast wasteland. Sponsors think controversy is unhealthy—for their business anyway." By not caving into this pressure, *CBS Reports* appeared to move beyond commercial concerns to place the public interest above pecuniary interest. When viewers tuned into the broadcast, then, they expected to find not only facts about pesticides, but facts that corporations had tried to hide from the public.[26]

CBS Reports promised to help audiences understand the debate generated by "the most controversial book of the year." While the program allowed for multiple responses, most TV critics scored the debate in Carson's favor. Many of them noted that throughout the show, Carson presented herself as reasonable and trustworthy while pesticides appeared menacing and destructive. As one columnist explained: "In a non-emotional voice . . . Carson read portions of her book. Cameras cut away to show billowing clouds of the chemicals being sprayed and later followed up with close-ups of the effects on birds, fish, and insects. Particularly devastating shots were those of a fish which appeared to be suffering from the equivalent of human lockjaw and a fallen bird rolling from side to side, tossing its wings in apparent agony." Critics suggested that Carson's dispassionate rhetorical style matched perfectly with the emotive response triggered by the visuals; together, the spoken commentary and images blended reason and emotion to convey Carson's critique of pesticide spraying programs.[27]

The mise-en-scène contrast between Carson and chemical industry spokesperson Robert White-Stevens accentuated the stark differences between their perspectives. While Carson often appeared close to nature, walking in the woods or near the sea (fig. 1.8), the interview with

FIGURE 1.8. Rachel Carson. Frame capture from *CBS Reports* television documentary "The Silent Spring of Rachel Carson," 1963.

FIGURE 1.9. Robert White-Stevens. Frame capture from *CBS Reports* television documentary "The Silent Spring of Rachel Carson," 1963.

White-Stevens was confined to only one setting: a laboratory, where he donned, quite appropriately, a white lab coat and spoke with confidence and authority about the salubrious effects of synthetic chemicals (fig. 1.9). The laboratory context, of course, lent an aura of legitimacy to White-Stevens but also signified an overly managed, lifeless world. Even as the lab coat offered a tangible visual marker of expertise and professionalism, it also activated a more disturbing symbol: the mad scientist, a figure from literary and filmic culture who represented "the isolation, the arrogance, the over-reaching hubris and the obsessive desire to tamper with things that are best left alone." While Carson appeared as a defender of fragile ecosystems, White-Stevens projected mad-scientist qualities, seeking to control nature for the benefit of humanity but also potentially gaining too much power to unleash destructive forces whose impact even experts could not fully understand or predict.[28]

Indeed, the repeated images of dying wildlife, coupled with the interviews of other scientists and government officials who admitted their ignorance about the long-term impact of pesticides, fostered feelings of doubt and uncertainty about the effects of toxic chemicals on the environment and human bodies. One critic noted, "On point after point, learned men conceded they have no answers to the vital, interrelated questions concerning the effects of pesticides on men and animals. It was a frightening demonstration of ignorance by men most intimate with the problem." Another commentator observed, "Man is rushing in to alter nature when he is ignorant of the end result of his activities. . . . What right have men to go ahead as if they knew the answer when they definitely do not?" The program thus challenged the authority of official experts and suggested that Carson was right to question the indiscriminate use of pesticides and to call for some kind of slowdown in the headlong rush to control nature.[29]

As Earth Day 1970 approached, Carson's paradigm of fear became more frequently depicted in the visual media. While *Life* and *National Geographic* had previously denied the hazards of DDT, these same magazines now visualized the threats to birds and to the larger ecological systems that sustained them. Popular images depicted a crisis of reproduction in which the severe thinning of eggshells prevented chicks from hatching. According to magazine articles and captions, as DDT became concentrated in the food chain it weakened the ability of bald eagles, ospreys, and other species to reproduce. Striking visual evidence of pesticide poisoning could be seen in the widely circulated images of cracked, damaged, and deformed eggshells. *National Geographic* featured a photograph of a dozen mallard eggs, many of which were "cracked or crushed by the mother" or otherwise "failed to develop." Similarly, *Life* showed a pair of ibis eggs—one cracked and crushed, the other dented in the middle—resting on an abandoned nest, a site completely devoid of living ibises (fig. 1.10).[30]

In response to this *Life* piece, one letter writer articulated the ecological lens through which many readers likely viewed these heartwrenching images. "Although the thought of a birdless world is a depressing one," she observed, "man's naïveté is even more frightening. From an ecological point of view man is just another species; the extinction of one species only brings closer the extinction of all species. When is man going to realize that the passing of a bird not only means a quiet, birdless sky but is a harbinger of things to come: today the ibis, tomorrow man?"[31]

Meanwhile, environmental groups sought to ban DDT and strategically deployed images as part of their campaign. In particular, the Environmental Defense Fund produced advertisements that represented DDT as posing a serious and escalating risk to humans, nature, and their conjoined futurity. Published in the *New York Times*, one ad provocatively asked: "IS MOTHER'S MILK FIT FOR HUMAN CONSUMPTION?"

FIGURE 1.10. Unhatched ibis eggs, damaged because of DDT pesticide poisoning on the Texas Gulf. Photograph by George Silk, 1970. Used by permission of Time & Life Pictures / Getty Images.

Below these large, boldface words, a photograph shows a white woman gently cradling a white infant. "Nobody knows," the text continues. "But if it were on the market it could be confiscated by the Food and Drug Administration. Why? Too much DDT." Another advertisement featured a photograph of a peregrine falcon and asked: "Have YOU seen a peregrine falcon lately? Or a bald eagle, osprey, brown pelican, . . . ? Chances are you haven't. Your grandchildren may NEVER see them." Together, these advertisements framed environmental citizenship as a long-term, future-oriented vision that included the right of people, especially children and those born in the future, to be protected from environmental harm and to live in a world graced by charismatic bird species.[32]

Soon, the name of DDT alone became a sign of crisis, encapsulating the sense of incremental doom that marked the temporal politics of environmentalism. In 1969, to mark its centennial, the American Museum of Natural History in New York City launched a controversial exhibit entitled *Can Man Survive?* For the next two years, the show delivered an apocalyptic warning about the environment to record-breaking audiences. The mass media, in its coverage of the exhibit, gave special attention to the museum's target audience: children and youth, along with their parents. *Life* magazine's review featured one image: a photograph of a young boy gazing at the montage of flickering lights, flashing images, and textual displays—including, in large, illuminated form, the letters "DDT." James Oliver, the museum's former director, penned a guest editorial about the exhibit for a popular parents' magazine, in which he outlined the apocalyptic tenor of modern life, "the signs of biological danger that are everywhere." Oliver believed that *Can Man Survive?* posed a question that all parents should consider: "What kind of world do we want our children to live in?" The three letters "DDT" projected the sense of gathering doom, the conjoined fate of people and nature in a world of escalating crisis.[33]

Television news coverage of the campaign to ban DDT emphasized the long-term risks the pesticide posed to both birds and people. These broadcasts featured pathos-inducing imagery of dead or dying songbirds and of pelican eggs too thin to hatch; they also discussed the threats to people, including infants who ingested DDT from their mother's milk. TV news reports explained how DDT persisted in the environment for many years and how it became increasingly concentrated in the food chain, endangering the health and survival of those at the top—from bald eagles to human beings. News commentators also routinely praised Rachel Carson for having first awakened the public to the dangers of pesticide use. Eric Sevareid explained on CBS in 1969, "What is happening is a growing shift to the premise of Miss Rachel Carson—that the benefit of the doubt must be given to life not death."[34]

While Carson's ecological ideas and feelings of doubt became increasingly accepted in American public culture, this coverage also narrowed the scope of *Silent Spring* to single out DDT as the nation's sole pesticide danger. The repeated emphasis on DDT, while lending legitimacy to the ban, also worked to marginalize systemic critiques of industrial agriculture and the increasing reliance on pesticides. Even as the visual media helped picture the long-term accumulative danger of DDT, the focus on this particular pesticide as an isolated problem made the ban appear to be a quick fix to the crisis. The slow violence of environmental time became more visible, yet the singular attention to DDT rendered invisible the ecological risks of other pesticides and of the broader system of corporate agriculture.

The first Earth Day celebration in 1970, the 1972 federal ban on DDT, and the emergence of the environmental regulatory state during this same period all demonstrated the profound shift in emotional style in

American public culture. As tools of persuasion and vectors of emotion, visual images played an active role in teaching Americans to see long-term environmental risk. Yet these images also indicated the limits of media representation: from the exclusive focus on atmospheric testing to the sole emphasis on DDT, the visual media obscured ongoing environmental dangers generated by nuclear weapons production and by the continued, escalating use of pesticides in agriculture.

These examples point to a number of recurring themes in the visual politics of American environmentalism. From SANE ads to the DDT ban, images helped popularize notions of the ecological body by explaining the ways in which strontium 90 and pesticides could enter the food chain and thereby threaten fragile ecosystems and human health. Like other images of environmental crisis that followed, they mobilized the private realm of home and family to depict children in danger and to call upon the state to protect the citizenry from harm. These images created a picture of universal victimhood, portraying all Americans, no matter where they lived, no matter their race or class, as being equally vulnerable to environmental dangers. Finally, these examples show how environmental thinkers and activists sought to inject long-term ecological concerns into American public culture. Even as these time frames of fear acquired legitimacy, though, the mainstream media often looked to immediate short-term solutions to these accretive dangers. The Limited Test Ban Treaty and the DDT ban appeared to resolve the problems raised by antinuclear activists and pesticide critics. In both cases the temporality of resolution seemed to displace the temporality of fear.

SANE ads, LBJ campaign commercials, and the use of DDT imagery all erased the lines between nature and culture to warn of a gradually escalating crisis that threatened both human health and ecological systems. In challenging the Cold War emotional style, these images depicted environmental danger as part of everyday reality and provided the visual backdrop to the rise of popular environmentalism.

PART ONE

Earth Day and the Visual Politics of Environmental Crisis

TWO

From Santa Barbara to Earth Day

In February 1969, eighteen-year old Kathy Morales walked along a beach in Santa Barbara, California. She paused, leaning down to stroke the feathers of an oil-soaked loon. After gently caressing the bird, she stood up and in frustration exclaimed: "He's just dying. Let him die in peace." As tears welled up in Kathy's eyes, the television news reporter asked what she could do to help this bird—along with the cormorants, seals, ruddy ducks, and other creatures that were suffocating and dying on the beach around her. "The only way is to stop the oil platform from killing off nature," she responded. The camera zoomed in, focusing closely on Kathy's face. She continued to cry and then ended the interview with a frightening prediction: "We will soon destroy ourselves."[1]

Only a few years earlier, the Beach Boys, in songs like "Surfin' USA," had celebrated the glistening sands of California as a landscape of teenage abandon where America's youth could find endless amusement along the shores of the Pacific. Now, as the 1960s came to a close, ABC News portrayed to a national audience a remarkably different vision: a teenager crying on a beach in Santa Barbara, mourning the loss of wildlife and glimpsing a scene of apocalyptic despair. Following Kathy's comment, the camera cut back to the loon, flapping its wings and still struggling on the sand. The reporter crouched down near the loon. "For years," he observed, "these beaches were the place for the good life—the parties, the picnics." The loon let out a haunting cry as the reporter concluded: "Now these sands are a wildlife graveyard."

A week before television brought Kathy's tears into millions of American homes, a Union Oil Company well ruptured off the coast of Santa Barbara. Even as oil leaked into the Pacific and formed a massive slick, early reports suggested that the beaches might be spared. Soon, however, the winds shifted, and oil began to cover the beaches and blacken the sands, spelling disaster for thousands of birds, seals, and other species of marine life. In the weeks and months that followed, oil spill imagery filled American television screens and popular magazines. Almost every report contrasted the black, oily crude with the white, shiny sand, the menacing presence of death with the ebullient sense of life that characterized the California coast.[2]

Throughout 1969 and beyond, the Santa Barbara oil spill became a defining symbol of the nation's environmental crisis. Like Kathy Morales, many other Americans would greet the grisly spectacle of death with sadness and even tears, would feel intense emotion at the sight of wildlife perishing on the Pacific shores. Like her, they would also wonder if the death of these creatures foretold more terrible times ahead. If the fate of loons and sea lions was intertwined with the fate of humanity, then perhaps Americans were on the verge of doomsday.

Although other environmental disasters would beset the American landscape in 1969, including the notorious fire on the Cuyahoga River in Cleveland, none would receive as much media attention as was given to the Santa Barbara oil spill. Unlike the Cuyahoga fire, which was extinguished even before photographers and television crews could arrive at the scene, the visible effects of the Santa Barbara spill continued for weeks to come, thus providing the media with an unfolding narrative to cover and ongoing opportunities to produce images of environmental devastation. The visual appeal of the story was heightened by California's place in the American imagination; the state was associated, particularly during the 1960s, with a sense of fresh hope and possibility, of new beginnings in an uncommonly beautiful setting. To see, as *Newsweek* magazine put it, this "normally serene Pacific-coast community" befouled by a "thick sludge that glistened like blackberry jam" vividly demonstrated the idea of universal vulnerability: if even the gleaming coast of California could fall prey to "the lethal muck," then presumably every place in the United States might be visited upon by some form of environmental calamity.[3]

Santa Barbara demonstrated how the long time frames of environmental crisis could now shape popular perceptions of even singular tragedies, such as this catastrophic oil spill. Coverage of this event thus constituted a crucial and surprising moment in the media's representation of environmental time. The conventions of media coverage—the focus on immediate visibility and spectacular violence—would lead us to expect this oil spill to be depicted as an isolated disaster, a cataclysmic, sensational event detached from the slowly escalating problems of the environmental crisis. Yet quite the opposite happened: Santa Barbara would be routinely cited by the media and audiences alike as emblematic of a broader long-term sense of ecological danger. This popular view was based not just on what Americans saw in the oil spill—not just on the actual, visible evidence of catastrophe—but on what they had been trained to see by SANE, Rachel Carson, and others who had helped bring the ecological lens into American public culture. The emotional specter of oiled wildlife fused with cognitive understandings of the imperiled ecological fabric to make this event a symbol of threatened futurity.

Santa Barbara produced powerful images of lost innocence: an affluent coastal city besieged by a black, poisonous tide; young sea lions rejected by their own mothers, left to starve after being tainted with the scent of petroleum; young women crying as they watched birds die, worrying that technology would soon unleash a killing spree against people as well. The oil spill registered such a passionate response and triggered so much emotion among Americans because it seemed both anomalous and typical: a place by the sea, formerly sheltered from the environmental problems that beset other cities, now engulfed by pollution, victimized by the undulating waves of oil. No longer immune, no longer set apart from the rest of the United States, Santa Barbara became a poster child for the ubiquitous spread of pollution and for the gnawing, unsettling sense that all of America would soon be overtaken by the environmental crisis. As the dark, contaminated waves washed upon the shore, as birds and other wildlife suffocated and died, as teardrops fell from young women's eyes, the enormity of the crisis sank in: the perception that everywhere was threatened, the feeling that everyone should worry about the future.

In December 1968, just six weeks before Kathy Morales cried on TV, the *Apollo 8* astronauts circled the moon and returned with the legendary *Earthrise* photograph. This iconic image depicts the earth aglow in colors of white, blue, and brown, surrounded by utter blackness, rising over the desolate lunar landscape (fig. 2.1). On the Christmas Day front page of the *New York Times*, the poet Archibald MacLeish argued that

FIGURE 2.1. *Earthrise*. Photograph by William Anders, 1968. Courtesy of NASA.

earth imagery would transform human perception: "To see the earth as it truly is, small and blue and beautiful in that eternal silence where it floats, is to see ourselves as riders on the earth together." Many environmentalists shared MacLeish's enthusiasm for the image. Believing that it represented the frailty, uniqueness, and interdependence of life on this planet, they latched onto the photograph as a symbol of their movement. "Apollo 8 . . . reminds us that the rivers, the seas, the creatures, the air of this beautiful, life-giving planet exist nowhere else in the emptiness of the universe we know," a letter writer to the *New York Times* explained. Similarly, the countercultural *Whole Earth Catalog* hailed *Earthrise* as a pivotal image that "established our planetary facthood and beauty and rareness (dry moon, barren space) and began to bend human consciousness."[4]

In the aftermath of the Santa Barbara oil spill, *Earthrise* and other earth images gained increasing visibility in American public culture and played a prominent role in the first celebration of Earth Day on April 22, 1970. Senator Gaylord Nelson of Wisconsin, who had been deeply affected by media imagery of the oil spill, developed the idea for Earth Day. In August 1969, Nelson visited Santa Barbara to see the damage for himself and began to contemplate "what could be done to bring public opinion to bear on the lethargic political community." He suggested that environmentalists should follow the example of the antiwar movement in organizing teach-ins around the nation—a series of meetings and discussions about the environmental crisis. Nelson's modest proposal tapped into popular anxiety over pollution and became a much larger and far more influential event than he had envisioned. Indeed, approximately twenty million Americans participated in Earth Day events around the nation, making it the largest protest in US history.[5]

A few days before Earth Day, the *New York Times* claimed the day should be considered "Space Day as well as Earth Day." "It is no mere coincidence that the great recent upsurge in ecology has coincided with the historic space triumphs at the end of the 1960s," the editors explained. "An easily comprehended visual perspective on earth's place in the universe was unobtainable until craft were available that could take cameras thousands of miles into deep space." Many Earth Day participants reiterated this idea. The distinguished anthropologist Margaret Mead, speaking to a crowd at Bryant Park in New York City, praised photographs of the earth taken from outer space. "We have today the knowledge and the tools to look at the whole earth," Mead observed. "I think the tenderness that lies in seeing the earth as small and lonely and blue is probably one of the most valuable things we have now."[6]

The popular meanings attached to earth imagery during this period revealed the shift in emotional styles in American public culture. While President John F. Kennedy had envisioned space exploration and the prospect of a moon landing as signs of American technological dominance on a new frontier, the environmentalist focus on the fragile earth instead emphasized ecological vulnerability and feelings of fear about the planetary future. While Kennedy's confident glimpse into this extraterrestrial frontier celebrated the onward march of progress, many Americans now saw in these pictures a poignant reminder about the accumulating dangers to their earthly home.[7]

From Santa Barbara to Earth Day and beyond, audiences encountered an array of images that signified ecological interconnection and the all-encompassing reality of the environmental crisis. While some images focused on particular places or spectacular disasters, the coverage of environmental crisis transcended specific locales to imagine Americans living through a critical moment, a temporal horizon in which the future looked increasingly uncertain and threatening. Yet, as we will see, the environmental icons that emerged during the period surrounding Earth Day also bequeathed troubling blind spots to

American environmental politics: most of the imagery emphasized the notion of universal vulnerability and thereby obscured the power relations that determine the unequal experience of environmental risk. In addition, even as Santa Barbara and other oil spills became associated with the accretive apocalypse described by Carson, the specific long-term dangers of oil dependency rarely figured into public discussions of these visually spectacular petro-calamities.

The visual media represented the Santa Barbara spill through panoramic portraits from above as well as through more intimate ground-level shots that focused on the plight of individual creatures. Aerial photography captured the scope of this disaster through wide-angle shots of the sea and shore. Steadily encroaching upon the placid shoreline, the oil appeared as an aggressive force, an enemy that advanced, one magazine caption observed, in "stringlike formations." The aerial pictures helped viewers imagine that they were looking down upon a battlefield, surveying the wreckage of war. These views reinforced the widespread use of military metaphors to describe the spill; the oil itself was often referred to as an invasive force. "Life in Santa Barbara today is somewhat reminiscent of civilian life in a war zone," the *New York Times Magazine* explained. "[The city] lives under constant threat of further oleaginous invasion from the sea."[8]

While the aerial views revealed the slick's enormous reach, the pictures of oil-drenched animals called for a more emotional response, and urged audiences to feel sadness and sympathy toward the death of helpless creatures. Almost all of the images that used this representational strategy focused on individual animals. Detached from other wildlife, many of these creatures appeared completely immobilized by oil. If a bird in flight offered a quintessential glimpse of wild nature, then birds coated in oil, rendered motionless by the imprisoning viscosity, represented a cruel perversion of the natural order. While petroleum enabled human society to enjoy speed and mobility, this same substance trapped nonhuman creatures in the human-induced muck.

These pictures also revealed the breakdown of natural processes, including, in perhaps the most poignant example, the utter collapse of the relationship between mothers and their offspring. One disturbing *Life* magazine article began with a photograph of an infant sea lion, his eyes fixed on the camera as he peered over a large rock on San Miguel Island, thirty-five miles from the ruptured oil well (fig. 2.2). The sea lion, "himself stained by the spillage," the caption noted, "surveys his threatened domain." *Life* informed readers that the young pup would

FIGURE 2.2. Infant sea lion following Santa Barbara oil spill. Photograph by Harry Benson published in *Life*, June 13, 1969. Used by permission of Harry Benson Ltd.

soon die of starvation: "The odor of the oil would camouflage his own distinctive scent so that his mother would consider him not her own and refuse to nurse him."[9]

By emphasizing the damaged relations between mothers and infants, the destruction of one of nature's most powerful bonds, *Life* sought to elicit an emotional response in readers: "Here and there . . . oil-drenched pups . . . cried weakly and thrashed about like scalded rats, their eyelids gummed shut, umbilicals stained and caked." Just as ABC News had shown Kathy Morales crying, *Life* described twenty-two-year-old Judy Smith, a visitor to San Miguel, "quietly weeping" as she pondered the situation. "'I feel as though I had gone into somebody's house where everyone was murdered, for no reason at all,'" she said. Her comment defined the oil spill not as a random accident, an aberration for which no one was responsible, but rather as a crime against nature that murdered the young.[10]

The spill made Union Oil appear particularly destructive and rapacious. The company's negative image would be exacerbated by a quote attributed to Fred Hartley, the company president, who spoke before a Senate subcommittee on the same day that Kathy Morales shed tears for the imperiled loon on the beach. Hartley callously remarked, "I'm amazed at the publicity for the loss of a few birds"—or at least so claimed many newspaper and magazine reports.[11]

Already reeling from the oil spill coverage, Union Oil worked to contain the fallout from this remark. The company took out a full-page advertisement in the *New York Times*. "Please, let's set the record straight," the ad began, and then featured a lengthy letter from Hartley explaining "that, at no time, anywhere, did I make that insensitive remark charged to me." (Versions of this letter would also appear in *Time*, the *Wall Street Journal*, and other publications.) Having received so much negative attention from the spill, company officials wanted to make sure that their president did not appear "insensitive" to the plight of nature. "The tragedy of the Santa Barbara off-shore oil well accident is serious enough without generating a greater emotional storm by charging me with a wholly false and fictitious statement," Hartley commented.[12]

While Hartley described public reaction to Santa Barbara as an "emotional storm," popular views of the spill instead combined cognition with emotion to present the event as a sign of the ongoing degradation of the environment. The popularization of the ecological lens led many observers toward Carson-inspired visions of slow-motion apocalypse: they described blackened beaches and oil-covered wildlife as evidence of the escalating dangers of the environmental crisis. As a sudden, unexpected event, the spill would seem to conform to conventions of media temporality, to appear as an isolated accident, nothing more than an unplanned environmental disaster. Indeed, in a special environmental issue of *Ramparts*, the leading New Left magazine, the sociologist Harvey Molotch critiqued media coverage of Santa Barbara for emphasizing the immediacy of events rather than the slow violence of environmental time. "The rather extensive media coverage of the oil spill centered on a few dramatic moments . . . such as the pathetic deaths of the sea birds struggling along the oil-covered sands," Molotch argued. According to him, the emotionality of the coverage limited public understanding of environmental problems.[13]

Yet Santa Barbara marked a pivotal moment in popularizing the concept of environmental crisis, a threatening, long-term condition that reached well beyond the shores of the Pacific to encompass the entire American landscape. The spill, many people claimed, was not an isolated incident that merely marred one stretch of the California coast, but one that instead represented the disturbing future of the American environment. *Life*'s story about the sea lions on San Miguel inspired many readers to write letters, including one who expressed his anxiety about the escalating crisis. "Your article . . . should frighten all men as much as the news of Hiroshima," he wrote. "The world itself is being threatened."[14]

Audiences looked beyond the visually spectacular example of Santa Barbara to glimpse other crises in the making, and to see this oil spill as a sign of the slow-motion violence of the environmental crisis. This perception involved the fusion of fact and feeling, science and spectacle: audiences responded emotionally to the plight of imperiled wildlife and filtered these feelings through the ecological lens that had already brought warnings of radioactive fallout, pesticide dangers, and other threats to humans and nature. As one letter writer explained to the *New York Times*, "The public rage is not just an outcry against the deaths of birds, however much these deaths grieve many, but an outcry against a symbol of what unrestrained development has produced. The deaths of birds, and seals, and dolphins, and fish may remind us that we too are threatened by the deadly air and water that we have contaminated." The pictures of dying birds fostered emotional identification with the fate of individual creatures, but also triggered a larger sense of concern, and mobilized public feelings about the escalating threats to the American environment.[15]

Yet even as Santa Barbara lent legitimacy to environmental time, the media and many environmentalists also detached the event from the specific context of energy policy and oil dependency in modern America. By seeing Santa Barbara as part of a general, all-encompassing environmental crisis, the particular problems associated with oil consumption receded from view. "Oil pollution was only one of a litany of perceived environmental disasters capturing headlines," the historian Paul Sabin explains. "As a result, many legislative victories that followed Santa Barbara broadly addressed environmental problems but had a less specific and tangible impact on the issue of oil." The numerous reforms adopted during this pivotal environmental moment, Sabin concludes, "did not fundamentally alter the country's underlying dependence on oil." While the oil spill appeared as a sign of the long-term incremental danger of environmental crisis, the delayed, accretive effects of oil dependency—including, we now know, the release of carbon dioxide into the atmosphere and the gradual warming of the planet—remained invisible.[16]

Near the end of 1970, for its special issue titled "Our Ecological Crisis," *National Geographic* used a picture from Santa Barbara—of a western grebe swimming through an oil slick—to represent this sense of an all-encompassing crisis (fig. 2.3). Appearing both on the magazine's cover and as part of the feature article, the image reveals patterns of color produced by the menacing presence of oil: glowing halos and shimmering, iridescent ripples surround the grebe. Despite its aesthetic qualities, the photograph also contains what the critic Roland Barthes once described as the "imperious sign . . . of future death." As the accompanying article noted, "The bird was doomed. "The oil so damaged his feathers that human help would be of no avail." The lone grebe seems to swim with purpose, yet readers know to see the photograph as a marker of doom, an ominous sign of the bird's impending death.[17]

National Geographic's special issue helped solidify the popular perception of Santa Barbara as a symbol of something much larger and more threatening than a single oil spill confined to a particular place: the meanings of the event spilled well beyond the California coast to signal a broader set of ecological concerns. *National Geographic* presented nature and human society as interlinked to suggest that the environmental crisis spelled disaster not just for this western grebe, but also, potentially, "to man's only home." With Santa Barbara prominently displayed on the cover, the feature article began with a photograph of the whole earth. After celebrating the planet's "life-support systems," the magazine warned: "But for centuries we have taken them for granted, considering their capacity limitless. . . . Unless we stop abusing our vital life-support systems, they will fail." Linking Santa Barbara to earth imagery, *National Geographic* presented the oil spill as an event that belonged not just to the news cycles but to the longer timescale of environmental crisis.[18]

From Santa Barbara to Earth Day, the environmental crisis became the conceptual frame through which the media portrayed pollution and other environmental problems. This coverage revealed how the long time frames of fear became increasingly accepted in American public culture, and how the shift in emotional styles made possible the emergence of popular environmentalism. By popularizing the ecological lens, media images helped produce an environmental public culture whose members learned to see themselves as potential victims of an escalating crisis.

During the period surrounding Earth Day 1970, other popular images encouraged spectators to grapple with fundamental questions: How

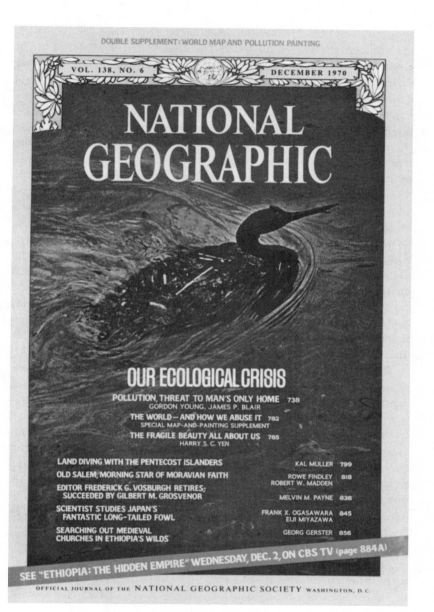

Cover text:

DOUBLE SUPPLEMENT: WORLD MAP AND POLLUTION PAINTING

VOL. 138, NO. 6   DECEMBER 1970

NATIONAL GEOGRAPHIC

OUR ECOLOGICAL CRISIS

SEE "ETHIOPIA: THE HIDDEN EMPIRE" WEDNESDAY, DEC. 2, ON CBS TV (page 884A)

OFFICIAL JOURNAL OF THE NATIONAL GEOGRAPHIC SOCIETY WASHINGTON, D.C.

FIGURE 2.3. *National Geographic* cover, December 1970. Photograph by Bruce Dale. Bruce Dale/National Geographic Creative.

were they—as individuals, families, and communities—affected by the environmental crisis? Who was responsible for the devastation of the environment—corporate elites, government officials, or each and every American? How could the crisis best be solved—through individual or collective means?

The following four chapters will look at four environmental icons—pictures of people wearing gas masks, the comic strip character Pogo, the Crying Indian, and the recycling logo—that crystallized a narrative of crisis and response. Images of gas masks personalized the sense of risk by showing the ecological threat intruding upon the daily lives of all Americans, warning that everyone, no matter their race or class, could suffer equally from the deadly spread of pollution. These images ignored the vastly unequal experience of environmental risk to promote a vision of universal vulnerability. This emphasis on the widespread danger of the crisis and its frightening entry into private life also worked to personalize the sense of responsibility. The visual media—particularly through Pogo, the Crying Indian, and the recycling logo—obscured larger social structures to suggest that Americans could ward off doomsday by altering their actions in daily life. Emphasizing the notion of universal responsibility, popular images framed individual action as the prime solution to environmental problems.

These images reveal a paradox embedded in the environmental politics of this period. From the founding of the Environmental Protection Agency to the adoption of new federal policies to clean the air and water, this period led to significant environmental reform and expressed confidence in the government's ability to solve social problems. Yet even as the environmental regulatory state expanded, the visual media emphasized individual responsibility. To end pollution, Americans were told to look both to the federal government and to their own actions in daily life. These seemingly opposing trends, however, actually reinforced one another. The visual media popularized the concept of environmental crisis but also narrowed the meanings of environmentalism, too often severing it from questions of power and presenting it as a moralistic cleanup crusade. During this pivotal environmental moment, popular images masked systemic causes and ecological inequities to celebrate environmentalism as a consensual cause that everyone could support.

THREE

Gas Masks: The Ecological Body under Assault

In January 1970, three months before the first celebration of Earth Day, *Life* magazine joined other popular periodicals in making environmentalism the focus of a feature article. The issue begins with an eerie, almost otherworldly image. A white woman walks down a street pushing a white child in a stroller: a simple portrait of everyday life, except that both the woman and the toddler are wearing gas masks (fig. 3.1). John Pekkanen, who wrote the story and whose two-year-old daughter Sarah is pictured in the image, described the sense of fear that gripped him as he researched the piece, a "feeling of dread about the prospects of my own two children" growing up in "'a world without a future.'" Pekkanen was "deeply shaken" by the dire warnings of leading scientists, who told him that unless Americans solved the problem of air pollution, "we would all be walking the streets in gas masks ten years from now." In the photograph the gas masks worn by Sarah and Lucy, the wife of photographer Mike Mauney, were meant, Pekkanen continued, "to be symbolic of what's ahead."[1]

The image resembles a family snapshot, a casual picture of daily life—yet it is also a vision of the apocalyptic future. The haunting quality of the image is accentuated by Lucy's appearance: wearing a stylish leather coat along with a striped scarf carefully draped around her neck, she seems to accept the need for a gas mask with ready equanimity. Rather than expressing fear, her face, clearly vis-

FIGURE 3.1. Sarah Pekkanen and Lucy Mauney wearing gas masks. Photograph by Michael Mauney published in *Life*, January 30, 1970. Used by permission of Time & Life Pictures / Getty Images.

ible through the oval-shaped mask, reveals the hint of a smile. Meanwhile, Sarah's mask calls to mind the bug-eyed devices worn by troops on the Western Front during World War I; obscuring her face, the ghoulish headgear turns Sarah into a disturbing symbol of the future awaiting America's children. With this photograph, *Life* magazine offered its readers a snapshot from the impending apocalypse.

John Pekkanen was not alone in seeing the gas mask as a suggestive emblem of the environmental crisis. Indeed, it became ubiquitous in the visual discourse of pollution. In a 1970 editorial cartoon, widely printed in environmental magazines as well as the popular media, Ray Osrin from the Cleveland *Plain Dealer* provided an updated version of Auguste Rodin's statue, *The Thinker* (fig. 3.2). While Rodin's figure

The Thinker

FIGURE 3.2. "The Thinker." Editorial cartoon by Ray Osrin originally published in the Cleveland *Plain-Dealer* and reprinted in the *New York Times*, January 25, 1970. Courtesy of Ray Osrin Estate.

seems sternly sober, his mind fixed on metaphysical matters, the face of Osrin's Thinker is shielded by a gas mask that protects him from the pollution surrounding his meditative perch. Environmental groups reproduced this cartoon on a poster and also circulated other posters featuring people in gas masks. In a particularly striking example that linked religious imagery to environmental concern, one poster shows Christ hanging from a smokestack cross above a littered landscape, speaking through his mask into the smoggy air: "Father forgive them for they know not what they do." In a lighter vein, *Mad* magazine presented its freckle-faced hero, Alfred E. Neuman, outfitted in a gas mask while reading a "special polluted issue" of the publication. On Earth Day itself, the mass media seized upon the gas mask as a defining symbol of environmental protest: numerous television and newspaper reports described participants in cities around the nation using the masks as theatrical props to convey their anger at the condition of the urban environment.[2]

Even more than the Santa Barbara oil spill, the gas mask became an environmental icon that burrowed into public consciousness and symbolized an all-encompassing crisis in the making. Consider the results of the "Dirty Pictures Contest," a 1970 art competition sponsored by *Psychology Today* magazine for works addressing "the causes and effects of dirty air, water and neighborhoods." When the judges announced the winners in the fall, associate editor Scot Morris reported that the response to the competition was "immediate and urgent—almost overwhelming." Organizers "had to rent a warehouse to display the thousands of entries," some from professional artists, others from people who "confessed they had never before drawn or taken pictures." Out of all the visual symbols represented in the entries, Morris observed, "gas masks were the most popular." Among all the ways in which Americans could imagine the environmental crisis, the gas mask seemed the most resonant image, the overwhelming choice of this eclectic mix of contestants who often attached it to other familiar symbols. Gas masks, Morris reported, "appeared on babies, on skulls, on Miss America, on the earth, and on the bald eagle. At least five persons put gas masks on the Statue of Liberty." The judges awarded first prize in the mixed-media category to a "six-inch medallion" that "looks like a quarter, but George Washington is wearing a gas mask." The coin's date appeared as "197?," asking whether the nation's environment would survive the new decade.[3]

The gas mask served as a visual representation of scientific concerns about human survival in an increasingly degraded environment. Morris Neiburger, a professor of meteorology at UCLA, carefully studied the smog that was accumulating with each passing year, hanging like a pall over his city. He concluded that it bespoke a menacing, slowly escalating form of catastrophe. His apocalyptic language would be frequently cited in the period leading up to Earth Day. "All civilization . . . will pass away," Neiburger warned, "not from a sudden cataclysm like a nuclear war, but from gradual suffocation in its own wastes." The gas mask warned of the accretive hazards to human bodies enmeshed in ecological systems. Like SANE ads and other texts that challenged the Cold War emotional style, this imagery relied on the ecological lens to emotionalize the slow-motion disaster of pollution.[4]

In visually symbolizing the threat to fragile ecological bodies, the gas mask implicitly endorsed the rights of citizens to be protected from harm and placed demands upon the state to expand its regulatory agenda. Antipollution groups began using the gas mask as a protest prop in the 1950s and 1960s in campaigns that called upon local governments to combat smog and clean the air. In many cases, middle-class white women spearheaded these organizations and grounded their protests in traditional gender ideals, suggesting that as mothers and homemakers they had entered the public sphere to protect their children. These activists believed that the gas mask could mobilize parental feelings by portraying the escalating risks to future generations. As Earth Day approached, the gas mask moved beyond the local context of air pollution activism to represent environmentalist claims that the federal government needed to broaden its regulatory powers to protect the health of the citizenry.[5]

For the most part, gas mask imagery focused on white Americans as archetypal members of the national community and therefore as universal emblems of environmental danger. The apparently placeless ubiquity of air pollution, together with the repeated focus on white bodies, obscured the specific geographies of environmental risk and the social processes that produced environmental injustice. The gas mask acted as a sign of universal danger, of the white ecological body under assault. The story of this environmental icon thus reveals the limits of mainstream imagery: even as the gas mask normalized fear of the environmental future, its depiction of universal vulnerability obscured the vastly unequal experience of environmental risk.

The gas mask took on environmental meanings during this time, but it had long been considered a symbol of death and destruction. Indeed, the power of this symbol derived in part from its prior claim on the cultural imagination. During World War I the mask protected soldiers from asphyxiating gas but also became a purveyor of pejorative meaning, a symbol of the dehumanizing effects of modern warfare. Gas masks robbed soldiers of their individual identities, made them feel like bug-eyed creatures trying to survive amid the poisoned clouds of war. "Gas took the war into the realm of the unreal, the make-believe," the historian Modris Eksteins argues. "When men donned their masks they lost all sign of humanity." Following World War I, peace activists in the United States used the gas mask as a symbol to warn of the cataclysm that another war might bring, especially the development of deadlier gases targeted at civilians. Not just soldiers but all people would be subject to the lethal effects of poison drifting through the air.[6]

In the post–World War II period, gas mask imagery would reappear, not to signify the terror of total war, but rather to represent the possibility of gradual ecological collapse. In 1954, in an event that likely marked the debut of the gas mask in environmental protest, a group of air pollution activists led by middle-class white women staged a march through Pasadena, California. Wearing gas masks and carrying placards that warned of the menace of smog, the women called themselves the "Smog-a-Tears," derived from Walt Disney's Mouseketeers, the stars of a children's TV program. The women, many of whom were housewives, used their status as nurturing mothers to emphasize the need to protect vulnerable children from the perils of pollution. Eight years before Dr. Spock and SANE ads presented children as emotional emblems of the environmental future, several kids participated in this march, wearing gas masks to signify the threats to their bodies. The youngest marcher, a three-year-old white girl named Agatha Acker, not only donned a gas mask but even outfitted her favorite doll in a miniature mask (fig. 3.3).[7]

The gas mask emerged as an environmental protest symbol in a particular setting: an affluent, suburban southern California community seeking to control the escalating problem of smog. Yet it soon moved beyond this specific location to become an environmental icon, a visual marker that transcended place. Just as Life's photograph of Sarah and Lucy suggests no place in particular—only a soft-focus, vaguely urban scene appears in the background—the gas mask would often circulate as a symbol detached from place, removed from the particularities of local conditions to represent the notion of universal vulnerability.

Throughout the 1960s, as air pollution gained increasing public attention, the gas mask continued to signify environmental anxiety. In 1965 the satirical songwriter Tom Lehrer proclaimed, "Pollution,

FIGURE 3.3. Agatha Acker wearing a gas mask and holding a doll at smog protest, Pasadena, California, 1954. Copyright Bettmann/Corbis.

pollution / Wear a gas mask and a veil / Then you can breathe long as you don't inhale." The following year, John Gardner, the secretary of health, education, and welfare under Lyndon Johnson, warned that the device might offer, in the very near future, the only means of protection in an increasingly polluted America. "Our choices are narrow," he explained. "We can remain indoors and live like moles for an unspecified number of days each year. We can issue gas masks to a large segment of the population." In 1967, three years before *Life*'s photograph, the paperback edition of Barry Commoner's *Science and Survival* featured a similar image—of a gas-mask-wearing mother pushing a gas-mask-wearing child in a stroller—on its cover. The widespread use of the gas mask as a symbol again demonstrates the shift in emotional styles in American public culture. In these examples, three men—a songwriter, a cabinet secretary, and a scientist—all deployed the object to announce their fear of the environmental future.[8]

As Earth Day 1970 approached, the gas mask would appear more and more frequently across a diverse array of visual media. Emanating from below, a protest strategy used by activists in Pasadena and elsewhere, the gas mask would become one of the most recognizable images of environmental crisis. As a symbol of pollution's placeless ubiquity, it raised the specter of universal vulnerability and promoted the rights of citizens to be protected from environmental harm.

The gendered claims of the Smog-a-Tears and other air pollution activists suggest another way to view *Life*'s photograph of Sarah and Lucy: as a secularized image of the Madonna and child, set not in the biblical past but in the apocalyptic future. Madonna images typically depict bodily closeness and motherly protection, but these features are not apparent in the *Life* photograph. Isolated in their masks, *Life*'s Madonna and child are unable to share the warmth and affection enjoyed by most parents and children.

Not just air pollution, but other environmental concerns—especially the presence of pesticides in breast milk—would be represented through reference to Madonna imagery. A series of environmental posters, including one reproduced in *Time* magazine two months before Earth Day, depicted the bare breasts of white women from various camera positions. One poster offers a close, detailed shot, focused only on the breast itself, while another shows a woman's hair draped over her breast, pointing toward her obviously pregnant belly (fig. 3.4). In each poster, the following language is printed over the naked breast: "Caution: Keep out of the reach of children." Text in another part of

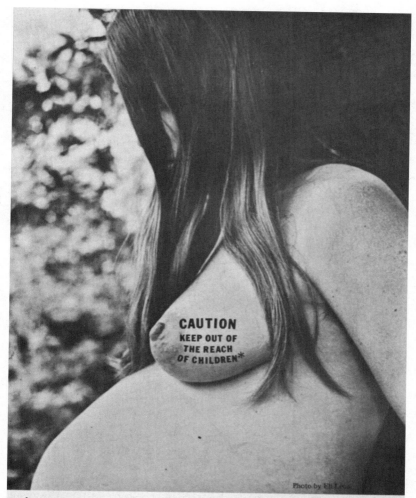

FIGURE 3.4. Ecology Center, *Caution: Keep Out of the Reach of Children*, 1969. Work on paper, 22.75 × 15.25 in., published in *Time*, February 2, 1970. Collection of the Oakland Museum of California, gift of the Rossman family.

the poster warns of the high DDT content in the milk of nursing mothers. Like the photograph of Sarah and Lucy, the posters convey fear about the world children are being born into. The posters link this concern to religious iconography by invoking the tradition of the nursing Madonna. In medieval art, the Virgin Mary was often portrayed breast-feeding Christ, her milk seen as "an emanation of heaven." In modern America, milk flowing from the breast instead appears as an unnatural poison, profaning the sacred connection between mother and infant.[9]

Throughout the twentieth century, secular portraits of the Madonna and child had frequently circulated in American visual culture. Dorothea Lange's *Migrant Mother* (1936), the most famous photograph of the Depression era, joined many other New Deal images in portraying a woman and her children as helpless victims of the depression. After World War II, photographers associated with the United Nations International Children's Emergency Fund (UNICEF) and other organizations drew on this tradition to generate concern for impoverished people in Africa and other parts of the global South. In these cases, the Madonna and child signified victims who deserved the sympathy of viewers, people who needed to be rescued, either through government assistance or donations offered by spectators. The images were directed at middle-class audiences; it was assumed that viewers did not suffer from the poverty and hunger that afflicted the subjects of the photographs.[10]

In striking contrast, environmental images did not differentiate the Madonna and child from viewing audiences, but rather suggested that subjects and spectators inhabited a shared geography of ecological risk. *Life*'s photograph evokes this theme of universal victimhood. Although Sarah and Lucy appear white and comfortably middle class, their gas-masked faces illustrate the way pollution endangers the health of all Americans. Likewise, posters of pregnant and nursing women did not portray any signs of poverty. Closely cropped, focusing on the woman's breast and pregnant belly with little evidence of the surrounding context, the images suggested that any woman, no matter her social or economic position, could pass deadly substances to her nursing infant.

Gas mask imagery and posters warning of pesticides in breast milk helped popularize the concept of the ecological body by demonstrating the link between human health and the condition of the surrounding environment. These pictures also provided the rationale for state action, making visible the need to regulate industry and ban particular pesticides. Yet lurking within this imagery was an unspoken assumption of racial privilege that helped create what the historian Michelle Murphy calls a "regime of imperceptibility." The same media that helped make the environmental crisis visible to a mass public also made invisible the various spaces in which different groups of Americans encountered pollution, pesticides, and other environmental threats. Even though some scientists, such as Barry Commoner, had argued that racialized minorities were "the special victims of pollution," subjected in their ambient environments, workplaces, and often substandard housing conditions to greater quantities of "smog, carbon monoxide, lead," and other pollutants, this imagery located all Americans within a common geography of environmental danger. The visual media heightened fears of the environmental crisis while ignoring the nation's widespread environmental inequities. Pictures encouraged audiences to perceive pollution as a generalized apocalyptic danger that affected all Americans equally, but also to turn their eyes away from the ecologies of injustice that marked the spaces in which different people lived and worked, the spaces that shaped their experience of the environmental crisis.[11]

Many Earth Day organizers wanted to forge connections with other social struggles. Rather than defining the environment as a separate entity, a crisis to be solved by pollution control measures, they sought to link the environmental cause to the fight against systematic racism and poverty in the United States. Three months before Earth Day, national coordinator Denis Hayes asserted that environmentalism meant far more than a cleanup campaign. "For ecology is concerned with the total system—not just the way it disposes of its garbage," he explained. "Our goal is not to clean the air while leaving slums and ghettos, nor is it to provide a healthy world for racial oppression and war." In the newsletter published by Earth Day organizers, the editors likewise sought to broaden the meanings of environmentalism by covering topics—including lead poisoning in the inner cities—that were typically ignored by mainstream conservation and environmental groups. "We've got rats and roaches, and we've got lead poisoning," the newsletter quoted one welfare rights activist as saying. "That's our environment." By calling attention to these issues, Earth Day organizers placed environmental hazards within a broader frame of social justice.[12]

This vision of environmentalism departed from the popular notion of universal vulnerability. While gas mask imagery and posters warning of pesticides in breast milk depicted all Americans as equally susceptible to environmental harm, other issues—including lead poisoning and farmworkers' exposure to pesticides—indicated the inequities of environmental risk. Yet media imagery, together with many mainstream environmental groups, promoted a narrow definition of the environment as a specific entity detached from social relations and structures of power. In promoting the concept of universal vulnerability, gas mask imagery popularized the environmental cause but also worked to narrow the scope of environmentalism, to sever it from issues of social justice.[13]

The notion of universal vulnerability obscured the racial and class dimensions of environmental risk and ignored other forms of environmental danger that primarily targeted the poor and racial minorities. In 1969 the journalist Jack Newfield eloquently explained why the accretive hazards of lead poisoning remained hidden from view. Even as the mainstream media increasingly publicized the escalating problems of the environmental crisis, lead poisoning "was a story hard to make visible or dramatic for the television networks. . . . How do you show a *process*, how do you show indifference, how do you show invisible, institutionalized injustice, in two minutes on Huntley-Brinkley? How do you induce the news department of a television network to get outraged about nameless black babies eating tenement paint . . . ?" During the time surrounding Earth Day, the visual media publicized the incremental dangers of the environmental crisis, but failed to consider the structures of power that produced environmental inequities. While white ecological bodies became signs of all-encompassing danger, the racially marked experience of environmental injustice rarely entered into mainstream framings of environmentalism.[14]

The mass media, moreover, found no place for African Americans in the environmental movement. Media coverage of Earth Day described the overwhelming whiteness of the event's participants and implied that pollution was irrelevant to the plight of minority groups. NBC News interviewed Michael Harris, an African American student at Howard University, who dismissed Earth Day as a "calculated political move by the established order in this country to divert attention from the pressing problems of black people." Other media reports emphasized similar criticisms of the environmental cause, including the pointed words of Mayor Richard G. Hatcher of Gary, Indiana, who argued that this newfound "concern with environment has done what George Wallace was unable to do: distract the nation from the human problems of the black and brown America, living in as much misery as ever."[15]

Even as the media provided some space for criticisms of Earth Day, this same coverage also ignored the efforts of subaltern communities to form an alternative vision of environmentalism. In St. Louis, for example, an organization called Black Survival performed a series of skits on Earth Day dramatizing the environmental problems of the inner city: high rates of air pollution that led to asthma, emphysema, and other respiratory ailments; inadequate city services, such as infrequent trash removal that resulted in rat and roach infestations; and a frightening epidemic of lead poisoning among children in the city's poorest neighborhoods. Black Survival grew out of a larger campaign against lead poisoning, a struggle coordinated by Ivory Perry, Freddie Mae Brown, and other civil rights activists in conjunction with scientists based at Barry Commoner's Center for the Biology of Natural Systems at Washington University. Like other environmental activists who sought to ensure the health of children, these St. Louis organizers stressed the accretive danger to young bodies. "Of all the housing problems Ivory Perry confronted, lead poisoning struck him as the most serious," the historian George Lipsitz explains. "It robbed children of their futures, deprived the community of the contributions those children might make, and operated so silently that no one even suspected its existence." Seeking to make the problem more visible, the Earth Day skits featured poignant moments, including a father learning that his baby has died of lead poisoning, and voiced radical sentiments, including chants of "Black power" and "Power to the people." Although newspapers in St. Louis described these skits and other actions, Black Survival went unmentioned in national media coverage of Earth Day.[16]

It is worth asking why this form of environmental theater received no media attention while white people wearing gas masks became central to mainstream views of the environmental cause. By enacting their position as an oppressed segment of the population, by revealing the particular environmental conditions of the inner city—problems, they emphasized, that were located within larger structures of power—the members of Black Survival posed an important challenge to the dominant image of American environmentalism. As the geographer Laura Pulido explains, subaltern groups, including African Americans in urban areas and Latino farmworkers, developed a form of environmental politics that situated "environmental concerns within the context of inequality and attempts to alter dominant power arrangements." In contrast, the environmental icons of this period promoted a vision of universal vulnerability: what the sociologist Ulrich Beck would later describe as a "generalized consciousness of affliction." This coverage obscured environmental hazards that were clearly marked by the unequal distribution of risk, and instead concentrated on problems that supposedly threatened all Americans equally.[17]

The sociologist Nathan Hare, writing in *Black Scholar* during the same month as the Earth Day celebration, reminded readers that the term "'ecology' was derived" from the Greek word *oikos*, "meaning 'house.'" Environmental activists had adopted the language of ecology, but Hare suggested that it seemed they had forgotten this etymology—or perhaps they were simply unwilling to concern themselves with the "household and neighborhood environment of blacks." Indeed, while Hayes and other Earth Day organizers sought to link environmentalism to social justice, many leading environmental groups did not join in the struggle against lead poisoning, and even refused to define it as an environmental problem. The journalist Jack Newfield, who described lead poisoning as "an environmental disease of the urban ghettos," suggested why the mass media and mainstream organizations neglected to confront this "silent epidemic." He wrote, "It seems that nothing is real to the media until it reaches the white middle class. . . . Then it is a crisis."[18]

The photograph of Sarah and Lucy represented the environmental crisis reaching the middle class, as did the posters that warned of DDT in breast milk. These pesticide images also became linked to another struggle, one that suggested the possibility of viable interaction between mainstream and subaltern movements. In 1969, as the United Farm Workers Organizing Committee (UFWOC) struggled to get union recognition and bargaining rights for its primarily Latino membership, the group also began to focus on farmworkers' exposure to pesticides. The UFWOC had earlier called for a grape boycott in order to link consumers to producers and put pressure on growers to accept its demands. As part of this boycott campaign, the UFWOC explained how grapes posed health threats to consumers as well as workers. The UFWOC thus suggested that all bodies were porous, that both consumers and workers could suffer from the "economic poisons" sprayed on grapes. In publicizing this dimension of the boycott, the UFWOC emphasized that nursing mothers could pass pesticides onto their babies. The vulnerable female body again became a central motif in environmental politics.[19]

The UFWOC viewed its antipesticide campaign as part of a larger struggle against the inequalities faced by Latino workers. Even though the UFWOC played an important role in removing DDT from the fields, the replacements for this pesticide—known as organophosphates—were in fact even more hazardous to workers (though not to wildlife). Nevertheless, most leading environmental organizations did not view the health of workers as an environmental issue and so, with few exceptions, did not consider the dangers of pesticide replacements. Although Hayes and other Earth Day organizers sought to broaden the meanings of environmentalism, lead poisoning and farmworker health rarely became part of the environmental agenda. By defining environmental problems in a narrow fashion, by focusing on technical solutions, and by refusing to ally itself with other social movements, mainstream environmentalism seemed to speak primarily for white, privileged Americans.[20]

Likewise, the federal environmental policies approved during this period assumed that "everyone breathes the same air and drinks the same water." These reforms sought to make sure that people like Sarah and Lucy would not have to wear gas masks in the future, but failed to consider the inequities of environmental risk. As one Earth Day organizer and critic of mainstream environmentalism explained, referencing the widespread use of the gas mask image: "If air pollution continues to get worse, the rich will produce the gas masks but they will not be the first to have to buy them."[21]

The generalized air and water pollution measures adopted during this period also created other problems: the proliferation of toxic waste dumps, located primarily in minority communities. "The primary legacy of the environmental movement and its resulting regulations was not so much a reduction of industrial waste but a transfer of wastes from water and air to the land," the historian Andrew Hurley writes. Because white privilege was inscribed into the American landscape, these apparently race-neutral policies in fact only worked to racialize the distribution of environmental risk even further, forcing "aggrieved racial minorities," as George Lipsitz explains, to "encounter higher levels of exposure to toxic substances." These new policies certainly helped clean up the nation's air and water, but they also exacerbated environmental inequities by increasing the levels of pollution in minority neighborhoods, the spaces hidden from view in media coverage of the environmental crisis.[22]

From the Smog-a-Tears to *Life*'s photograph of Sarah and Lucy, many gas mask images deployed traditional conceptions of gender to make women and children appear especially vulnerable to pollution. Yet men also wore gas masks in air pollution protests, suggesting that their bodies, too, were permeable and threatened. Much like "Dr. Spock Is Worried" and Osrin's gas-masked version of *The Thinker*, these male protesters rejected the masculine ideal of strength and invincibility to express anxiety about pollution. Appropriating the gas mask in their campaigns, environmental protesters—men, women, and children—activated the familiar association of the device with apocalyptic fears to issue a warning that all Americans could become victims of the air they breathed.[23]

Indeed, the gas mask image that circulated most frequently at the time, and which has been reproduced most often in Earth Day retrospectives ever since, featured a young white man wearing a mask. The picture was snapped by an Associated Press (AP) photographer at an Earth Day protest in New York City (fig. 3.5). At lunch hour that day, Peter Hallerman, a sophomore at Pace College, joined approximately thirty other students in a lunch-hour demonstration at City Hall Park. While the group represented a tiny gathering compared with the massive crowds that assembled elsewhere in the city, the students hoped, the writer Timothy Dumas explains, that "the City Hall press corps would straggle out to gather a bit of Earth Day color." An AP photographer saw Hallerman wearing a World War II–era gas mask and decided to orchestrate an image that would juxtapose the beauty of spring with the eeriness of the gas mask. He asked Hallerman to stand "in front

FIGURE 3.5. Pace College student in gas mask, Earth Day, New York City, April 22, 1970. Used by permission of AP Photo.

of a blossoming magnolia tree" and then lean over so that his mask "touched the pink-white blossoms."[24]

The photograph isolates the young man and the blossoming branches from the surrounding landscape to focus primarily on the disturbing interaction between mask and flower. The published image almost never included a caption bearing Hallerman's name; his anonymity reinforced the idea of universal vulnerability, suggesting that not any single person or any particular group but rather everyone could suffer from the environmental crisis. The profile view reveals his white skin, again signifying universal risk. A familiar symbol of renewal, of the promise of new life in the spring, the flower appears in this image to inhabit a denatured world, a time when the pleasurable scents of the season can safely be apprehended only by breathing through protective gear. Unlike the "Daisy Girl" commercial and unlike Sarah and Lucy in the *Life* photograph, the presence of Hallerman extends the apocalyptic warning: The environmental crisis means that not only women and children, putatively the most vulnerable members of society, but even young white men must cope with the hazards of an increasingly polluted world.

If gas mask imagery provided spectators with an answer to a key question of the time—How were they affected by pollution?—other images would attempt to answer a second question: Who was responsible for the environmental crisis? The media's emphasis on universal vulnerability would be paralleled by the repeated insistence upon universal culpability, the notion that all Americans were responsible for causing the environmental crisis. While gas mask pictures implicitly called for government action to protect ecological bodies, these images would instead encourage Americans to become proper environmental citizens by feeling pangs of guilt for all their polluting ways.

FOUR

Pogo: "We Have Met the Enemy and He Is Us"

On February 5, 1970, Roy Alexander, a biology teacher from Charlotte, North Carolina, was desperately searching for a comic strip. Inspired by a recent speech given by Senator Gaylord Nelson, a longtime environmental advocate and the founder of Earth Day, Alexander noted that Nelson had quoted the protagonist in Walt Kelly's popular comic strip *Pogo*: "'We have met the enemy and [he] is us.'" Unable to find the particular strip containing this statement, Alexander wrote to Kelly, explaining how he hoped to use the cartoon to promote an upcoming environmental conference. "As Senator Nelson pointed out," Alexander explained, Pogo's statement "quite vividly applies to the real roots of the environmental crisis facing us today." Alexander closed by asking Kelly to "provide us with a cartoon depicting this idea."[1]

One day later, Jack Alford, the chairman of the Department of Engineering at Harvey Mudd College, penned a similar letter to Kelly: "When people all around me are blaming each other for the state of our environment," Alford wrote, "it helps my sanity to recall the wise words of Pogo, 'We have met the enemy and he is us.' If we can spread this radical doctrine, we might even be able to start improving the situation. Is it possible to obtain a reproduction of the cartoon in which this statement appeared?"[2]

Before the first celebration of Earth Day, it would be impossible for Kelly to provide Alexander, Alford, or anyone else with copies of the strip featuring the statement—for none yet existed. Alexander and Alford, it turned out, were looking for a comic strip that had not yet been drawn. Despite what Nelson said, Kelly had not yet applied these words to the environmental crisis—though a version of the statement had originally appeared seventeen years earlier, as part of Kelly's effort to criticize Senator Joseph McCarthy's anticommunist crusade. Beginning in 1969, the words were deployed again—first as part of an environmental exhibit at the American Museum of Natural History and then by a wide array of journalists, activists, and politicians. In writing to Kelly, Alexander and Alford expressed a longing to see, in visual form, what many Americans would come to believe was the most perceptive insight into the causes of the environmental crisis.

Walt Kelly soon obliged. Just in time for the first Earth Day, he created a poster in which Pogo the Possum, with anxious eyes and a quizzical brow, stands beneath a pair of stately, flowering trees (fig. 4.1). The landscape before him is covered with litter as numerous bottles, cans, and other detritus fill the space. Pogo turns his head back to look directly at the viewer and deliver his famous words: "We have met the enemy and he is us."[3]

This statement—not the words of a leading ecologist or political activist, but rather a sentence uttered by a cartoon marsupial—would become the most frequently repeated refrain to describe the environmental crisis. Although detached from any particular image, a free-floating statement that expressed the idea of human responsibility for environmental devastation, the words would be placed in dialogue with a variety of pictures to offer mixed-media commentaries on the environmental crisis. Pogo's words became so familiar and so widely accepted that Kelly's illustrations—both the Earth Day poster and subsequent strips featuring the quote—proved to be superfluous. His words elicited feelings of guilt, prompting many to join the pensive opossum in condemning all Americans for causing the environmental crisis.

Why did so many Americans embrace this statement? What did they discover in these words about the environmental crisis, the question of responsibility, and the meanings of environmental citizenship? The popular enthusiastic response to Pogo complicates any simple reading of dominant media institutions manipulating a passive audience into accepting a particular vision of environmental culpability. Pogo's words—and the images associated with them—did not merely reflect or neatly follow broader shifts in US public policy and political-economic practices. Indeed, his focus on personal responsibility preceded the individualist turn in American public culture and the subsequent

FIGURE 4.1. "We Have Met the Enemy." Earth Day poster by Walt Kelly, 1970. Copyright Okefenokee Glee & Perloo Inc. Used by permission.

triumph of neoliberal values and structures. During this pivotal moment in the history of environmentalism, Pogo helped Americans reimagine their political world. Even as the federal government adopted a broad range of new policies to protect the citizenry from environmental harm, Pogo and other cultural texts projected the idea of "governing at a distance," of individual citizens engaging in new practices of self-regulation to combat environmental problems.[4]

The story of Pogo's statement, considered by comic strip scholars to be "the most quoted sentence in the history of the American comic strip," is a story, at least in part, of a sentence trying to find a proper visual home, of words that perhaps do not even need an image to convey their message. Yet this statement would have never entered the environmental lexicon without the prior (and ongoing) significance of *Pogo* to American political culture. The appeal of *Pogo*—a comic strip that used visual and verbal wordplay to promote liberal values—offers an important context in which to view the statement's origin and its later incarnation as the most widely circulated commentary on the environmental crisis. Walt Kelly's politics, together with the critiques sparked by the statement's popularity, also reveal the possibilities and limits of environmentalism during the period surrounding Earth Day. In the end, the statement found not one but multiple visual referents, and resonated with a plethora of media images and texts that emphasized personal responsibility for the environmental crisis.[5]

Pogo helped popularize environmental guilt, making this emotion central to mainstream framings of the environmental cause. In the aftermath of Earth Day, as Walt Kelly continued to promote this vision of environmentalism, some environmental activists sought to distance themselves from Pogo by countering: "We have met the enemy and he is *some* of us." Emphasizing power relations, these activists questioned the politics of guilt and rejected the emotional framework of environmentalism propagated by Pogo and the visual media. While gas masks conveyed the idea of universal vulnerability, Pogo evoked the notion of universal responsibility, a perspective, these activists claimed, that ignored the systemic causes of the environmental crisis and constrained the meanings of environmental citizenship. Kelly's political satire, they believed, had become nothing more than social satire, a depoliticized form of address that encouraged Americans to laugh at themselves rather than acting to improve the environmental future.[6]

During the 1950s, in hundreds of newspapers, Walt Kelly skewered Joseph McCarthy for his abrogation of constitutional principles and his merciless, unwarranted attacks on individual citizens. Kelly's treatment of political matters, together with the strip's imaginative landscape populated with memorable characters (including Albert Alligator, Deacon Mushrat, Porkypine, and Pogo the Possum), endeared *Pogo* to an educated, liberal readership who appreciated its effort to elevate mass culture by bringing thoughtful, informed commentary into a comic strip. Launched in 1948, *Pogo* went into syndication the following year, and by the end of the 1950s it appeared in approximately five hundred newspapers around the world, boasting an audience of about fifty million readers. Described as "an idol among the eggheads," *Pogo* would be eagerly followed by liberal Americans who marveled at Kelly's ability to use humor as a political weapon, to puncture conservative pieties through verbal and visual satire. While political cartoonists pilloried McCarthy on the editorial page, Kelly did so in the funny pages, a domain that in the 1950s was usually devoid of politics. Indeed, *Pogo* was alone among popular comic strips in addressing controversial subjects like the rise of McCarthy.[7]

Kelly did not blame McCarthy alone for the dastardly events of McCarthyism, and he believed that the American people had allowed a demagogue to exploit their anxieties. To make this point, he revised a statement made by Commodore Oliver Hazard Perry, who, after defeating the British at the Battle of Lake Erie during the War of 1812, exclaimed: "We have met the enemy and they are ours." In his introduction to a 1953 *Pogo* anthology, Kelly tweaked this remark. "We shall meet the enemy," he wrote, "and not only may he be ours, he may be us." According to Kelly, the enemy was not just one senator but an entire nation that acquiesced.[8]

Before Senator Nelson quoted Pogo, the statement appeared in the *Can Man Survive?* exhibition, which ran at the American Museum of Natural History from 1969 to 1971. Described by the *New York Times* as "a mixed-media show of psychedelic pandemonium," *Can Man Survive?* sought to immerse audiences in a different way of experiencing history, to make them see the past as a story of decline rather than of progress. Like other depictions of environmental crisis, *Can Man Survive?* evoked the ecological lens to warn of the accretive, escalating dangers to both humans and nature.[9]

Can Man Survive? shocked many viewers. The critic Wilfrid Sheed, reviewing the show for *Life* magazine, explained how it confounded his expectations, made him *feel* much differently than he usually felt in a natural history museum. "One of the gentler memories of any

New Yorker's childhood is the Sunday visit to the American Museum of Natural History," Sheed remarked. "There . . . you could see how evolution had finally worked out: rickety, loose-knit dinosaurs, stuffed grizzlies, and You, the Lord of Creation. One departed with the feeling that Nature's fangs were permanently under glass and Mankind was sitting pretty." *Can Man Survive?*, however, offered a jarring contrast to this experience: "The message . . . is that the way things are going, there may not even *be* any natural history much longer. When the old museum worries, it is probably time for everyone to worry." Both the medium and the message, the style and the substance, were meant to jolt audiences. "You can tell something is wrong the moment you walk in," Sheed reported. "A harsh, raised structure, appropriately called a truss, has been plunked down in the middle of sedate Roosevelt Memorial Hall. Inside the truss, walls jab at you, electronic music jangles you, ramps rise and ceilings dip—more like a fun-house than a biological cathedral."[10]

Can Man Survive? marked the debut of Kelly's quotation on an environmental stage. At the show, his words were strategically placed above a graph documenting the dramatic rise of world population: from a gradual, barely perceptible increase in the eighteenth and nineteenth centuries to a sudden surge in the twentieth. As the line stretched into the future it became almost completely vertical, suggesting that population numbers would continue to rise exponentially. Together with the exhibit's multimedia displays of pollution, pesticides, and other forms of environmental danger, the graph projected a negative image of futurity to warn of the escalating risk of the environmental crisis.[11]

During this period, concerns about population growth would also be represented on a widely distributed poster titled "Population Explosion," that showed human beings literally falling off the face of an overcrowded earth. When placed in dialogue with *Can Man Survive?* and other popular images, Pogo's statement seemed to confirm the message of the scientist Paul Ehrlich, whose best-selling *The Population Bomb* (1968) identified overpopulation as the root cause of all environmental problems. "The causal chain of the deterioration is easily followed to its sources," Ehrlich argued. "Too many cars, too many factories, too much detergent, too much pesticides . . . too little water, too much carbon dioxide—all can be traced easily to *too many people*."[12]

While *Can Man Survive?* specifically linked Pogo to the theory of overpopulation, many reviewers claimed that the quote transcended this particular issue to encapsulate the larger message of the show. One critic observed: "The aim [of the exhibit] is to evoke an emotional reaction against man's exploitation of the environment. The theme is taken from the cartoon strip Pogo." Another explained: "The shock of this confrontation with the enemy, which turns out to be us, seemed horrible in the extreme."[13]

At the end of the exhibit, a sign posted on the wall asked, "CAN MAN SURVIVE?" Audiences heard a recording of voices chanting, "It's up to you. It's up to you. It's up to you. . . ." This mantra would be repeated a final time in a different form, one that moved from the electronic multimedia environment of the show to a much simpler device. "You turn away, to leave," one reviewer noted, "and face . . . yourself, reflected in a mirror. Above you a sign reiterates simply, 'It's up to you.' And it is, you see." After leading spectators through its declension narrative, after bringing them to the edge of environmental ruin, the museum focused on the individual, reflected back with an image of the self. As Pogo's quote migrated from *Can Man Survive?* to other media sources, it would frequently be used in a similar fashion: to frame the issue of human survival as a question of personal responsibility.[14]

During the year 1970, Pogo's statement cropped up in many other sources. In January, for its lead story in a special issue on "The Ravaged Environment," *Newsweek* claimed that one of the main goals of environmentalists was to convince the American public that all were complicit in the current crisis. "For the villain of the piece is not some profit-hungry industrialist who can be fined into submission, nor some lax public official who can be replaced," the magazine explained. "The villains are consumers who demand . . . new, more, faster, bigger, cheaper playthings without counting the cost in a dirtier, smellier, sicklier world." According to this analysis, all participants in consumer culture—especially those who demanded and desired more "playthings"—shared equal responsibility for ravaging the nation's environment. Throughout its special issue, *Newsweek* tended to obscure the differences among different groups of Americans, ignoring issues of class, race, or power to suggest, like Pogo, that pollution resulted not from the nefarious actions of a single individual but rather from the accumulation of choices made by an entire nation of consumers: the enemy was us.[15]

Newsweek reinforced this message by pairing photographs with didactic captions. In one photograph, a person is seen walking down a city street murky with smoke. The caption reads: "Man has always been a messy animal." Linking air pollution to human nature, this comment naturalizes the environmental crisis, suggesting that "man"—and not any specific interest or group of people—deserves censure for ruining the natural world. Another page displays a collage of four images—a heap of junked automobiles, an aerial view of suburban sprawl, a pile of tin cans, and a large crowd of people—to draw a visual connection between litter, pollution, and population growth. "Too many people living too close together pile high the earth with worn-out junk and trash," the caption explains. Through these visual and verbal cues, *Newsweek* repeatedly echoed what would become Pogo's view of the environmental crisis. No wonder the author of the lead story chose to end his piece with Kelly's familiar refrain.[16]

As Earth Day approached, Pogo's words enjoyed increasing currency. In March a political columnist for the *Saturday Review* noted, with no apparent surprise, that a member of the White House Council on Environmental Quality—an advisory body appointed by President Richard Nixon—had quoted the cartoon character at a recent meeting. Pogo's statement echoed through the corridors of government office buildings in Washington as well as through the crowded streets of New York City, where hundreds of thousands gathered to celebrate Earth Day. As they walked around Union Square, the site of many Earth Day events, New Yorkers saw this statement emblazoned across a prominently displayed booth and printed on lapel buttons that were distributed by eager volunteers throughout the day. In the Earth Day edition of the *New York Times*, a book reviewer complained that leading environmentalists had not cogently explained the meanings of the environmental crisis. So he concluded as so many others did, by quoting the environmental thinker he found most sagacious: Pogo the Possum.[17]

Following Earth Day, Pogo's prominence continued to rise. In July, Wallace Stegner, the respected novelist and longtime member of the Sierra Club, reviewed a selection of environmentalist books in the pages of *Life* magazine, including two that tended, he complained, "to misrepresent the crisis as a we-they conflict." Stegner chose instead to "acknowledge Pogo's wisdom" by labeling the environmental crisis "a we-us conflict" for which all people were to blame. Pogo's statement even became the title for Stegner's essay. To illustrate his point, the piece featured a drawing by Walt Kelly—not one of Pogo, but rather an updated rendering of a Bible story. Surrounded by a zebra, a giraffe, and other creatures, Noah looks warily from the top of the ark as it founders on the shoals of a throwaway society, a conglomeration of tin cans and other discarded items (fig. 4.2). The quotation seems equally at home in this biblical setting, as it suggests a sinful state of humanity that lays waste to the sanctity of creation. Likewise, Stegner fused the ecological

FIGURE 4.2. Illustration by Walt Kelly published in *Life*, July 10, 1970. Copyright Okefenokee Glee & Perloo Inc. Used by permission.

lens with moral and religious rhetoric to argue that Americans shared a common experience of sin and victimhood. "There is only one ecology," he wrote, "and man is part of it—properly as member, too often as destroyer, ultimately as victim."[18]

As 1970 drew to a close, *National Geographic* published its special issue titled "Our Ecological Crisis" (which, as we saw in fig. 2.3, featured a western grebe swimming through Santa Barbara's oil slick). The editors reinforced Stegner's view, both by quoting Pogo admiringly and by including a series of photographs with captions pointing toward a similar conclusion. One image portrays a trash-strewn alley, wedged between tenement buildings in New York City. The photograph offers a closeup view of two dolls—one with white skin, the other with dark skin—atop the heap of rubbish, while a narrow slit of sky appears above, the only sign of "nature" amid this crowded urban landscape (fig. 4.3). Like Pogo and Stegner, the caption encourages viewers to locate this image within a universal framework of ecological sin: "Culprit and victim—man himself," it announces in boldface. The caption goes on to indict Americans for consuming too many resources and producing too much waste, thus turning the photograph into a mirror of blame.[19]

Although *National Geographic* noted that the photograph was taken near 143rd Street, the magazine failed to identify the neighborhood—Harlem, if it were West 143rd Street, the South Bronx if it were East—or to comment on environmental problems—including inadequate trash collection, rat and roach infestations, and lead poisoning—faced by the primarily African American and Latino populations in these areas. Harlem and South Bronx activists challenged these conditions, yet their subaltern vision of environmentalism did not fit with *National Geographic*'s generalized portrait of the environmental crisis. Indeed, the dark- and white-skinned dolls in the photograph provide the magazine's only suggestion of racial differences within the United States, and even here the photograph implies black-and-white unity rather

FIGURE 4.3. Trash in alley, New York City. Photograph by James P. Blair published in *National Geographic*, December 1970. James P. Blair / National Geographic Creative.

than division. Ignoring power relations and disregarding the experience of environmental injustice, *National Geographic* recited Pogo's words to explain the cause of "our ecological crisis."[20]

While many Americans interpreted Pogo's statement as a commentary on pollution and the environmental crisis, Kelly, in both his Noah's Ark illustration and his Earth Day poster, emphasized the presence of litter—the unsightly debris of consumer culture—rather than pesticides, pollution, and other environmental hazards. In these images we do not see oil-soaked wildlife or thinning eggshells or Pogo donning a gas mask. Instead, the pictures condemn the hideousness of litter and encourage viewers to take personal responsibility for their trash. Kelly's attack on consumer culture thus focused not on the broader environmental impact of increasing resource extraction, toxics production, and other unsustainable practices, but rather on individuals who carelessly disposed of their waste.

Pogo joined other media texts—including, as we will see in the next chapter, the antilitter campaigns of Keep America Beautiful—in imagining politics in an individualist frame by stressing the personal dimensions of environmental citizenship. At the very moment that the state expanded its role to protect the citizenry from environmental danger, Pogo and other media images disaggregated environmentalism into a set of private acts. Even as the long time frames of fear became normalized, the rhetoric of personal responsibility offered the prospect of a quick fix. The long-term, gradually escalating sense of danger might be overcome by the short-term, immediate actions taken by individuals who embraced Pogo's pronouncement. Long before neoliberal models of citizenship became dominant in American public culture, Pogo and other media texts conflated litter with pollution to personalize environmental politics and ignore the larger structures of power that produced the slow violence of the environmental crisis.[21]

In the time leading up to Earth Day, Kelly began to bristle at the rhetoric of some environmental activists, especially their claim that corporations and the government were primarily to blame for environmental degradation. Even though Kelly had long been a darling of the liberals, and even as he continued to mock Richard Nixon, Spiro Agnew, and other Republican politicians in his comic strip, he also believed that left-leaning critics of corporations were casting aspersions rather than acknowledging their own role in the environmental crisis. Just as McCarthy had tried to pin all the nation's problems on communist subversion, some leaders of the ecology movement, Kelly worried, purposely denied their own environmental sins to focus on the polluting actions of powerful organizations.

Kelly felt so strongly about this issue that he repeated his message many times—first in the poster of Pogo looking back from the pile of refuse, then in a number of strips published between 1970 and 1972, and finally in a book and never-completed film. On the first anniversary of Earth Day in 1971, in a strip featuring the quote, Kelly portrayed Porkypine, normally a grumpy character, waxing romantic as he walked through the forest. "Ah, Pogo," he says, "the beauty of the forest primeval gets me in the heart" (fig. 4.4). Pogo responds, for reasons that seem unclear at first: "It gets *me* in the feet, Porkypine." In the next panel, as the two characters gaze at an enormous heap of garbage, Porkypine remarks, "It *is* hard walkin' on this stuff." Pogo then delivers his famous words to explain the blight that surrounds them. In another strip, Albert Alligator, who has been "broodin' about pollution" for a long time, carelessly tosses his cigar into a bucket of lemonade, leading Porkypine to utter the familiar statement.[22]

In addition to his poster and strips, Kelly also wanted to produce an animated television special in which Pogo and his mates would muse about pollution. Although he died before finishing the film, Kelly published a book that featured drawings and dialogue based on the cartoon in progress. He began the book—titled *We Have Met the Enemy and He Is Us*—with a brief discussion of pollution. "The big polluter did not start out with smokestacks," Kelly wrote. "He didn't start out pumping gunk into the waters of our world when he was six years old. He started small. Throwing papers underfoot in the streets, heaving old bottles into vacant lots, leaving the remnants of a picnic in the fields and woodlands. Just like the rest of us." Kelly equated throwing trash on the side of the road with spewing poisons into the air and water. The litterbug and the corporate polluter, in his view, were brothers under the same skin.[23]

Pogo's statement tapped into an important strand of US environmentalism—the emphasis placed upon individual action. The mass-market paperbacks produced for the first Earth Day all featured a section that offered "consumer hints"—from carrying a lunch box to buying low-phosphate laundry detergents—to help solve the environmental crisis. Environmental activists focused on questions of individual behavior to encourage Americans to recognize the impact of consumption and to consider how their daily lives were enmeshed in broader ecological systems.[24]

FIGURE 4.4. *Pogo* comic strip by Walt Kelly, 1971. Copyright Okefenokee Glee & Perloo Inc. Used by permission.

Even though the question of individual responsibility formed an important part of their political vision, many of these same activists, including some Earth Day organizers, began to challenge the media and corporations for appropriating their rhetoric in a manner that ultimately obscured questions of power. "While [the capitalists] have been trying to fight off the environmental lawyers," one critic observed, "their PR men have been working overtime promulgating 'the Pogo syndrome.' It is called that because its spokesmen, who are with surprising frequency the vice-presidents of oil companies, frequently quote Pogo:

'We have met the enemy and he is us.' In other words, it's all your fault, however rich or poor you may happen to be. . . . The trouble with this argument is that it deludes people into thinking that their individual decisions can help. . . . Saving cans and bottles and newspapers on an individual basis and hauling them to 'recycling centers' may make you feel better; it does not help. . . . [It] is, in bleak fact, not worth a damn."[25]

Likewise, Bruce Kennedy, an activist from California, worried that his fellow environmentalists too often embraced the ersatz wisdom of Pogo. "Environmentalists who stress the value of individual action frequently cite the statement taken from the comic strip," he wrote. "As the statement implies, the major cause of pollution is people, their carelessness and wastefulness. . . . I believe that this conclusion is largely false, and more harm than good will come from an uncritical belief in the concept of environmental salvation through self-improvement. . . . The collective agents rather than the private individuals are mainly responsible for today's massive environmental pollution. . . . We must look outward to those collective agents rather than inward to our collective selves." Pogo, as Kennedy noted, encouraged Americans to look inward and feel guilty for the current state of the environment, to hold themselves and their careless and wasteful ways responsible for pollution. The idea of environmentalism as salvation through self-improvement threatened to deflect attention from the power of collective agents and to focus instead, relentlessly and obsessively, on the individual self.[26]

Kennedy did not completely reject the idea of individual responsibility, nor did he consider the environmentalist project as an either/ or dichotomy between personal and collective action. He noted that "do-it-yourself ecology" fostered "a badly needed ecological consciousness." "Indeed, individual action should be encouraged," he continued, "but only so long as its limitations are fully understood. . . . . To believe that individual self-improvement can solve the pollution problem is, at best, a delusion and, at worst, a cop-out. Widespread adoption of this approach would divert our energies from substantive activities to a preoccupation with minutiae, and reduce the growing environmental movement to a toothless utopianism." Kennedy believed that the guilt-inducing message of Pogo not only activated the wrong emotion but also created confusion over the question of responsibility and channeled environmentalist energies into a depoliticized quest for individual salvation.[27]

As Americans learned to see the environmental crisis, as they learned to imagine themselves living through a critical moment in time on the cusp of an even more alarming future, they also began to focus on understanding how their everyday actions made them complicit. To a certain extent Pogo, in holding human beings responsible for environmental devastation, voiced an oppositional identity that challenged the wastefulness of consumer culture. Yet in showing Pogo and his friends joking about their role in causing the pollution that marred their fictional swamp, Kelly also used social satire to ridicule human society in general, to mock human foibles rather than marshal political critique. The comic strip historian Robert C. Harvey clearly summarizes the differences between these satirical forms: "Social satire . . . ridicules homo sapiens generally—as a species or as a social creature—launching many a penetrating sidelong glance askance at the often silly conventions of society itself. Social satire induces us to laugh at ourselves. Political satire, on the other hand, aims at specific targets—malefactors and fools in public office, government policy gone awry, . . . and so on. Political satire aims at getting us to act as well as to laugh (and when we laugh, it is not at ourselves)." By adopting the mode of social satire rather than political satire, Kelly encouraged audiences not only to laugh, but also to look inward and feel a sense of guilt, perhaps even to alter their actions in daily life. Many environmentalists worried, though, that the laughter could just as easily foster a sense of complacency by making human devastation of the environment seem inevitable and unavoidable, part of the natural order of things.[28]

While gas masks and Santa Barbara demonstrated the all-encompassing danger of the environmental crisis, Pogo raised another question that the visual media would attempt to answer during the period surrounding Earth Day: How could the crisis best be solved—through individual or collective means? As we will see, a single tear rolling down the face of the Crying Indian would provide the iconic answer to that question, and would extend Pogo's effort to popularize environmental guilt.

FIVE

The Crying Indian

It may be the most famous tear in American history. Iron Eyes Cody, an actor in native garb, paddles a birch bark canoe on water that seems at first tranquil and pristine but becomes increasingly polluted along his journey. He pulls his boat from the water and walks toward a bustling freeway. As the lone Indian ponders the polluted landscape and stares at vehicles streaming by, a passenger hurls a paper bag out a car window. The bag bursts on the ground, scattering fast-food wrappers all over his beaded moccasins. In a stern voice, the narrator comments: "Some people have a deep abiding respect for the natural beauty that was once this country. And some people don't." The camera zooms in closely on Iron Eyes Cody's face to reveal a single tear falling, ever so slowly, down his cheek (fig. 5.1).

This tear made its television debut in 1971 at the close of a public service advertisement for the antilitter organization Keep America Beautiful. Appearing in languid motion on television, the tear would also circulate in other visual forms, stilled on billboards and print media advertisements to become a frame stopped in time, forever fixing the image of Iron Eyes Cody as the Crying Indian. Garnering many advertising accolades, including two Clio Awards, and still ranked as one of the best commercials of all time, the Crying Indian spot enjoyed tremendous airtime during the 1970s, allowing it to gain, in advertising lingo, billions of "household impressions" and achieve one of the highest viewer recognition rates in television history. After being remade multiple times to support Keep America Beautiful, and after becoming

FIGURE 5.1. The Crying Indian. Advertising Council / Keep America Beautiful advertisement, 1971. Courtesy of Ad Council Archives, University of Illinois, record series 13/2/203.

indelibly etched into American public culture, the commercial has more recently been spoofed by various television shows, including *The Simpsons* (always a reliable index of popular culture resonance), *King of the Hill*, and *Penn & Teller: Bullshit*. These parodies—together with the widely publicized reports that Iron Eyes Cody was actually born Espera De Corti, an Italian-American who literally played Indian in both his life and onscreen—may make it difficult to view the commercial with the same degree of moral seriousness it sought to convey to spectators at the time. Yet to appreciate the commercial's significance, to situate Cody's tear within its historical moment, we need to consider why so many viewers believed that the spot represented an image of pure feeling captured by the camera. As the television scholar Robert Thompson explains: "The tear was such an iconic moment. . . . Once you saw it, it was unforgettable. It was like nothing else on television. As such, it stood out in all the clutter we saw in the early 70s."[1]

As a moment of intense emotional expression, Iron Eyes Cody's tear compressed and concatenated an array of historical myths, cultural narratives, and political debates about native peoples and progress, technology and modernity, the environment and the question of responsibility. It reached back into the past to critique the present; it celebrated the ecological virtue of the Indian and condemned visual signs of pollution, especially the heedless practices of the litterbug. It turned his crying into a moment of visual eloquence, one that drew upon countercultural currents but also deflected the radical ideas of environmental, indigenous, and other protest groups.[2]

At one level, this visual eloquence came from the tear itself, which tapped into a legacy of romanticism rekindled by the counterculture. As the writer Tom Lutz explains in his history of crying, the Romantics enshrined the body as "the seal of truth," the authentic bearer of sincere emotion. "To say that tears have a meaning greater than any words is to suggest that truth somehow resides in the body," he argues. "For [Romantic authors], crying is superior to words as a form of communication because our bodies, uncorrupted by culture or society, are naturally truthful, and tears are the most essential form of speech for this idealized body."[3]

Rather than being an example of uncontrolled weeping, the single tear shed by Iron Eyes Cody also contributed to its visual power, a moment readily aestheticized and easily reproduced, a drop poised forever on his cheek, seemingly suspended in perpetuity. Cody himself grasped how emotions and aesthetics became intertwined in the commercial. "The final result was better than anybody expected," he noted in his autobiography. "In fact, some people who had been working on the project were moved to tears just reviewing the edited version. It was apparent we had something of a 60-second work of art on our hands." The aestheticizing of his tear yielded emotional eloquence; the tear seemed to express sincerity, an authentic record of feeling and experience. Art and reality merged to offer an emotional critique of the environmental crisis.[4]

That the tear trickled down the leathered face of a Native American (or at least someone reputed to be indigenous) made its emotionality that much more poignant, its critique that much more palpable. By designing the commercial around the imagined experience of a native person, someone who appears to have journeyed out of the past to survey the current landscape, Keep America Beautiful (KAB) incorporated the counterculture's embrace of Indianness as a marker of oppositional identity.

Yet KAB, composed of leading beverage and packaging corporations and staunchly opposed to many environmental initiatives, sought to interiorize the environmentalist critique of progress, to make individual viewers feel guilty and responsible for the degraded environment. Deflecting the question of responsibility away from corporations and placing it entirely in the realm of individual action, the commercial castigated spectators for their environmental sins but concealed the role of industry in polluting the landscape. A ghost from the past, someone who returns to haunt the contemporary American imagination, the Crying Indian evoked national guilt for the environmental crisis but also worked to erase the presence of actual Indians from the landscape. Even as Red Power became a potent organizing force, KAB conjured a spectral Indian to represent the native experience, a ghost whose melancholy presence mobilized guilt but masked ongoing colonialism, whose troubling visitation encouraged viewers to feel responsible but to forget history. Signifying resistance and secreting urgency, his single tear glossed over power to generate a false sense of personal blame. For all its implied sincerity, many environmentalists would come to see the tear as phony and politically problematic, the liquid conclusion to a sham campaign orchestrated by corporate America.

Before KAB appropriated Indianness by making Iron Eyes Cody into a popular environmental symbol, the group had promoted a similar message of individual responsibility through its previous antilitter campaigns. Founded in 1951 by the American Can Company and the Owens-Illinois Glass Company, a corporate roster that later included the likes of Coca-Cola and the Dixie Cup Company, KAB gained the support of the Advertising Council, the nation's preeminent public service advertising organization. Best known for creating Smokey Bear and the slogan "Only You Can Prevent Forest Fires" for the US Forest Service, the Ad Council applied the same focus on individual responsibility to its KAB advertising.[5]

The Ad Council's campaigns for KAB framed litter as a visual crime against landscape beauty and an affront to citizenship values. David F. Beard, a KAB leader and the director of advertising for Reynolds Metals Company, described the litter problem in feverish tones and sought to infuse the issue with a sense of crisis. "During this summer and fall, all media will participate in an accelerated campaign to help to curb the massive defacement of the nation by thoughtless and careless people," he wrote in 1961. "The bad habits of littering can be changed only by making all citizens aware of their responsibilities to keep our public places as clean as they do their own homes." The KAB fact sheet distributed to media outlets heightened this rhetoric of urgency by describing litter as an infringement upon the rights of American citizens who "derive much pleasure and recreation from their beautiful outdoors. . . . Yet their enjoyment of the natural and man-made attractions of our grand landscape is everywhere marred by the litter which careless people leave in their wake." "The mountain of refuse keeps growing," draining public coffers for continual cleanup and even posing "a menace to life and health," the Ad Council concluded.[6]

And why had this litter crisis emerged? The Ad Council acknowledged that "more and more products" were now "wrapped and packaged in containers of paper, metal and other materials"—the very same disposable containers that were manufactured, marketed, and used by the very same companies that had founded and directed KAB. Yet rather than critique the proliferation of disposables, rather than question the corporate decisions that led to the widespread use of these materials, KAB and the Ad Council singled out "individual thoughtlessness" as "the outstanding factor in the litter nuisance."[7]

Each year Beard's rhetoric became increasingly alarmist as he began to describe the antilitter effort as the moral equivalent of war. "THE LITTERBUGS ARE ON THE LOOSE," he warned newspapers around the nation, "and we're counting on you to take up arms against them. . . . Your newspaper is a big gun in the battle against thoughtless littering." Each year the campaign adopted new visuals to illustrate the tag line: "Bit by bit . . . every litter bit hurts." "This year we are taking a realistic approach to the litter problem, using before-and-after photographs to illustrate our campaign theme," Beard reported in 1963. "We think you'll agree that these ads pack a real wallop." These images showed a white family or a group of white teenagers enjoying themselves in one photograph but leaving behind unsightly debris in the next. The pictures focused exclusively on places of leisure—beaches, parks, and lakes—to depict these recreational environments as spaces treasured by white middle-class Americans, the archetypal members of the national community. The fight against litter thus appeared as a patriotic effort to protect the beauty of public spaces and to reaffirm the rights and responsibilities of citizenship, especially among the social group considered to exemplify the American way of life.[8]

In 1964, though, Beard announced a shift in strategy. Rather than appealing to citizenship values in general, KAB would target parents in particular by deploying images of children to appeal to their emotions. "This year we are . . . reminding the adult that whenever he strews litter he is remiss in setting a good example for the kids—an appeal which should hit . . . with more emotional force than appealing primarily to his citizenship," he wrote. The campaign against litter thus packaged itself as a form of emotional citizenship. Situating private feelings within public spaces, KAB urged fathers and mothers to see littering as a sign of poor parenting: "The good citizenship habits you want your children to have go overboard when they see you toss litter away."[9]

These new advertisements featured Susan Spotless, a young white girl who always wore a white dress—completely spotless, of course—together with white shoes, white socks, and a white headband. In the ads, Susan pointed her accusatory finger at pieces of trash heedlessly dropped by her parents (fig. 5.2). The goal of this campaign, Beard explained, was "to dramatize the message that 'Keeping America Beautiful' is a family affair'"—a concept that would later be applied not just to litter, but to the entire environmental crisis. Susan Spotless introduced a moral gaze into the discourse on litter, a gaze that used the wagging finger of a child to condemn individual adults for being bad parents, irresponsible citizens, and unpatriotic Americans. She played the part of a child who not only had a vested interest in the future but also appealed to private feelings to instruct her parents how to be better citizens. Launched in 1964, the same year that the Lyndon Johnson campaign broadcast the "Daisy Girl" ad, the Susan Spotless campaign also represented a young white girl as an emblem of futurity to promote citizenship ideals.[10]

Throughout the 1960s and beyond, the Ad Council and KAB continued to present children as emotional symbols of the antilitter agenda. An ad from the late 1960s depicted a chalkboard with children's antilitter sentiments scrawled across it: "Litter is not pretty. Litter is not healthy. Litter is not clean. Litter is not American." What all these campaigns assumed was a sense of shared American values and a faith that the United States was fundamentally a good society. The ads did not attempt to mobilize resistant images or question dominant narratives of nationalism. KAB did not in any way attempt to appeal to the social movements and gathering spirit of protest that marked the 1960s.[11]

With this background history in mind, the Crying Indian campaign appears far stranger, a surprising turn for the antilitter movement. KAB suddenly moved from its rather bland admonishments about litter to encompass a broader view of pollution and the environmental crisis. Within a few years it had shifted from Susan Spotless to the Crying Indian. Rather than signaling its commitment to environmentalism, though, this new representational strategy indicated KAB's fear of the environmental movement.

"Daddy, you forgot...every litter bit hurts!"

It happens in the best of families! Dad takes the kids out fishing and forgets that every litter bit hurts . . . in more ways than one! ■ Floating litter is a safety hazard. Litter causes pollution of waterways. Litter causes good fishing spots to be closed down . . . spoils the natural beauty of America's parks and recreation areas. And that's not the worst of it! ■ The good citizenship habits you want your children to have go overboard when they see you toss litter away. So, Dad, Mom, everybody—lead the way to the litterbag. Carry one in your boat and car. On camping trips, take litter out with you. Make it a family project to Keep America Beautiful!

SUSAN SPOTLESS SAYS
KEEP AMERICA BEAUTIFUL ©1964,
Keep America Beautiful, Inc.

Published as a public service in cooperation with The Advertising Council.

KEEP AMERICA BEAUTIFUL—MAGAZINE AD NO. KAB-208-64-—7" x 10" (100 SCREEN)
Agency: Dancer-Fitzgerald-Sample, Inc.—Coordinator: David F. Beard, Reynolds Metals Company

FIGURE 5.2. "Daddy, you forgot . . . every litter bit hurts!" Advertising Council / Keep America Beautiful advertisement, 1964. Courtesy of Ad Council Archives, University of Illinois, record series 13/2/207.

The soft drink and packaging industries—composed of the same companies that led KAB—viewed the rise of environmentalism with considerable trepidation. Three weeks before the first Earth Day, the National Soft Drink Association (NSDA) distributed a detailed memo to its members, warning that "any bottling company" could be targeted by demonstrators hoping to create an "attention-getting scene." The memo explained that in March, as part of a "'dress rehearsal'" for Earth Day, University of Michigan students had protested at a soft drink plant by dumping a huge pile of nonreturnable bottles and cans on company grounds. Similar stunts, the memo cautioned, might be replicated across the nation on Earth Day.[12]

And, indeed, many environmental demonstrations staged during the week surrounding Earth Day focused on the issue of throwaway containers. All these protests held industry—not consumers—responsible for the proliferation of disposable items that wasted natural resources and created a solid waste crisis. In Atlanta, for example, the week culminated with an "Ecology Trek"—featuring a pickup truck full of bottles and cans—to the Coca-Cola company headquarters. FBI surveillance agents, posted at fifty locations around the United States to monitor the potential presence of radicals at Earth Day events, noted that in most cases the bottling plants were ready for the demonstrators. Indeed, the plant managers heeded the memo's advice: they not only had speeches prepared and "trash receptacles set up" for the bottles and cans hauled by participants, but also offered free soft drinks to the demonstrators. At these protests, environmental activists raised serious questions about consumer culture and the ecological effects of disposable packaging. In response, industry leaders in Atlanta and elsewhere announced, in effect: "Let them drink Coke."[13]

The NSDA memo combined snideness with grudging respect to emphasize the significance of environmentalism and to warn about its potential impact on their industry: If legions of consumers imbibed the environmentalist message, would their sales and profits diminish? "Those who are protesting, although many may be only semi-informed, have a legitimate concern for the environment they will inherit," the memo commented. "From a business point of view, the protestors . . . represent the growing numbers of today's and tomorrow's soft drink consumers. An industry whose product sales are based on enjoyment of life must be concerned about ecological problems." Placed on the defensive by Earth Day, the industry recognized that it needed to formulate a more proactive public relations effort.[14]

KAB and the Ad Council would devise the symbolic solution that soft drink and packaging industries craved: the image of the Crying Indian. The conceptual brilliance of the ad stemmed from its ability to incorporate elements of the countercultural and environmentalist critique of progress into its overall vision in order to offer the public a resistant narrative that simultaneously deflected attention from industry practices. When Iron Eyes Cody paddled his birch bark canoe out of the recesses of the imagined past, when his tear registered shock at the polluted present, he tapped into a broader current of protest and, as the ad's designers knew quite well, entered a cultural milieu already populated by other Ecological Indians.

In 1967 *Life* magazine ran a cover story titled "Rediscovery of the Redman," which emphasized how certain notions of Indianness were becoming central to countercultural identity. Native Americans, the article claimed, were currently "being discovered again—by the hippies. . . . Viewing the dispossessed Indian as America's original dropout, and convinced that he has deeper spiritual values than the rest of society, hippies have taken to wearing his costume and horning in on his customs." Even as the article revealed how the counterculture trivialized native culture by extracting symbols of imagined Indianness, it also indicated how the image of the Indian could be deployed as part of an oppositional identity to question dominant values.[15]

While *Life* stressed the material and pharmaceutical accoutrements the counterculture ascribed to Indianness—from beads and headbands to marijuana and LSD—other media sources noted how many countercultural rebels found ecological meaning in native practices. In 1969, as part of a special issue devoted to the environmental crisis, *Look* magazine profiled the poet Gary Snyder, whose work enjoyed a large following among the counterculture. Photographed in the nude as he held his smiling young child above his head and sat along a riverbank, Snyder looked like the archetypal natural man, someone who had found freedom in nature, far away from the constraints and corruptions of modern culture. In a brief statement to the magazine he evoked frontier mythology to contrast the failures of the cowboy with the virtues of the Indian. "We've got to leave the cowboys behind," Snyder said. "We've got to become natives of this land, join the Indians and recapture America."[16]

Although the image of the Ecological Indian grew out of longstanding traditions in American culture, it circulated with particular intensity during the late 1960s and early 1970s. A 1969 poster distributed by activists in Berkeley, California, who wanted to protect "People's

Park" as a communal garden, features a picture of Geronimo, the legendary Apache resistance fighter, armed with a rifle. The accompanying text contrasts the Indians' reverence for the land with the greed of white men who turned the space into a parking lot. Likewise, a few weeks before Earth Day, the *New York Times Magazine* reported on Ecology Action, a Berkeley-based group. The author was particularly struck by one image that appeared in the group's office. "After getting past the sign at the door, the visitor is confronted with a large poster of a noble, if somewhat apprehensive, Indian. The first Americans have become the culture heroes of the ecology movement." Native Americans had become symbolically important to the movement, because, one of Ecology Action's leaders explained, "'the Indians lived in harmony with this country and they had a reverence for the things they depended on.'"[17]

Hollywood soon followed suit. The 1970 revisionist Western *Little Big Man*, one of the most popular films of the era, portrayed Great Plains Indians living in harmony with their environment, respecting the majestic herds of bison that filled the landscape. While Indians killed the animals only for subsistence, whites indiscriminately slaughtered the creatures for profit, leaving their carcasses behind to amass, in one memorable scene, enormous columns of skins for the market. One film critic noted that "the ominous theme is the invincible brutality of the white man, the end of 'natural' life in America."[18]

In creating the image of the Crying Indian, KAB practiced a sly form of propaganda. Since the corporations behind the campaign never publicized their involvement, audiences assumed that KAB was a disinterested party. KAB documents, though, reveal the level of duplicity in the campaign. Disingenuous in joining the ecology bandwagon, KAB excelled in the art of deception. It promoted an ideology without seeming ideological; it sought to counter the claims of a political movement without itself seeming political. The Crying Indian, with its creative appropriation of countercultural resistance, provided the guilt-inducing tear KAB needed to propagandize without seeming propagandistic.

Soon after the first Earth Day, Marsteller agreed to serve as the volunteer ad agency for a campaign whose explicit purpose was to broaden the KAB message beyond litter to encompass pollution and the environmental crisis. Acutely aware of the stakes of the ideological struggle, Marsteller's vice president explained to the Ad Council how he hoped the campaign would battle the ideas of environmentalists—ideas, he feared, that were becoming too widely accepted by the American public. "The problem . . . was the attitude and the thinking of individual Americans," he claimed. "They considered everyone else but themselves as polluters. Also, they never correlated pollution with litter. . . . The 'mind-set' of the public had to be overcome. The objective of the advertising, therefore, would be to show that polluters are people—no matter where they are, in industry or on a picnic." While this comment may have exaggerated the extent to which the American public held industry and industry alone responsible for environmental problems (witness the popularity of the Pogo quotation), it revealed the anxiety felt by corporate leaders who saw the environmentalist insurgency as a possible threat to their control over the means of production.[19]

As outlined by the Marsteller vice president, the new KAB advertising campaign would seek to accomplish the following ideological objectives: It would conflate litter with pollution, making the problems seem indistinguishable from one another; it would interiorize the sense of blame and responsibility, making viewers feel guilty for their own individual actions; it would generalize and universalize with abandon, making all people appear equally complicit in causing pollution and the environmental crisis. While the campaign would still sometimes rely on images of young white children, images that conveyed futurity to condemn the current crisis, the Crying Indian offered instead an image of the past returning to the haunt the present.

Before becoming the Crying Indian, Iron Eyes Cody had performed in numerous Hollywood films, all in roles that embodied the stereotypical, albeit contradictory, characteristics attributed to cinematic Indians. Depending on the part, he could be solemn and stoic or crazed and bloodthirsty; most of all, though, in all these films he appeared locked in the past, a visual relic of the time before Indians, according to frontier myth, had vanished from the continent.[20]

The Crying Indian ad took the dominant mythology as prologue; it assumed that audiences would know the plotlines of progress and disappearance and would imagine its prehistoric protagonist suddenly entering the contemporary moment of 1971. In the spot, the time-traveling Indian paddles his canoe out of the pristine past. His long black braids and feather, his buckskin jacket and beaded moccasins—all signal his pastness, his inability to engage with modernity. He is an anachronism who does not belong in the picture.[21]

The spectral Indian becomes an emblem of protest, a phantomlike figure whose untainted ways allow him to embody native ecological wisdom and to critique the destructive forces of progress. He confronts viewers with his mournful stare, challenging them to atone for their environmental sins. Although he has glimpsed various signs of pollution, it is the final careless act—the one passenger who flings trash at his feet—that leads him to cry. At the moment the tear appears, the narrator, in a baritone voice, intones: "People start pollution. People can stop it." The Crying Indian does not speak. The voice-over sternly confirms his tearful judgment and articulates what the silent Indian cannot say: Industry and public policy are not to blame, because individual people cause pollution. The resistant narrative becomes incorporated into KAB's propaganda effort. His tear tries to alter the public's "mind-set," to deflect attention away from KAB's corporate sponsors by making individual Americans feel culpable for the environmental crisis.

Iron Eyes Cody became a spectral Indian at the same moment that actual Indians occupied Alcatraz Island—located, ironically enough, in San Francisco Bay, the same body of water in which the Crying Indian was paddling his canoe. As the ad was being filmed, native activists on nearby Alcatraz were presenting themselves not as past-tense Indians but as coeval citizens laying claim to the abandoned island. For almost two years—from late 1969 through mid-1971, a period that overlapped with both the filming and release of the Crying Indian commercial—they demanded that the US government cede control of the island. The Alcatraz activists, composed mostly of urban Indian college students, called themselves the "Indians of All Tribes" to express a vision of pan-Indian unity—an idea also expressed by the American Indian Movement (AIM) and the struggle for Red Power. On Alcatraz they hoped to create several centers, including an ecological center that would promote "an Indian view of nature—that man should live *with* the land and not simply *on* it."[22]

While the Crying Indian was a ghost in the media machine, the Alcatraz activists sought to challenge the legacies of colonialism and contest contemporary injustices—to address, in other words, the realities of native lives erased by the anachronistic Indians who typically populated Hollywood film. "The Alcatraz news stories are somewhat shocking to non-Indians," the Indian author and activist Vine Deloria Jr. explained a few months after the occupation began. "It is difficult for most Americans to comprehend that there still exists a living community of nearly one million Indians in this country. For many people, Indians have become a species of movie actor periodically dispatched to the Happy Hunting Grounds by John Wayne on the 'Late, Late Show.'" The Indians on Alcatraz, Deloria believed, could advance native issues and also potentially teach the United States how to establish a more sustainable relationship with the land. "Non-Indian society has created a monstrosity of a culture where . . . the sun can never break through the smog," he wrote. "It just seems to a lot of Indians that this continent was a lot better off when we were running it." While the Crying Indian and Deloria both upheld the notion of native ecological wisdom, they did so in diametrically opposed ways. Iron Eyes Cody's tear, ineffectual and irrelevant to contemporary Indian lives, evoked only the idea of Indianness, a static symbol for polluting moderns to emulate. In contrast, the burgeoning Red Power movement demonstrated that native peoples would not be consigned to the past, and would not act merely as screens on which whites could project their guilt and desire.[23]

A few weeks after the Crying Indian debuted on TV, the Indians of All Tribes were removed from Alcatraz. Iron Eyes Cody, meanwhile, repeatedly staked out a political position quite different from that of AIM, whose activists protested and picketed one of his films for its stereotypical and demeaning depictions of native characters. Still playing Indian in real life, Cody chastised the group for its radicalism. "The American Indian Movement (AIM) has some good people in it, and I know them," he later wrote in his autobiography. "But, while the disruptions it has instigated helped put the Indians on the world map, its values and direction must change. AIM must work at encouraging Indians to work within the system if we've to really improve our lives. If that sounds 'Uncle Tom,' so be it. I'm a realist, damn it! The buffalo are never coming back." Iron Eyes Cody, the prehistoric ghost, the past-tense ecological Indian, disingenuously condemned AIM for failing to engage with modernity and longing for a pristine past when buffalo roamed the continent.[24]

Even as AIM sought to organize and empower Indian peoples to improve present conditions, the Crying Indian appears completely powerless, unable to challenge white domination. In the commercial, all he can do is lament the land his people lost.

The advertisement, though, does offer a prescription for action, expressed in the narrator's closing words: "People start pollution. People can stop it." Just as Walt Kelly portrayed Pogo picking up trash in the Earth Day poster (see fig. 4.1), KAB suggested that each individual could play a role in cleaning up the nation's environment. When the Crying Indian commercial first aired, NBC News described the group's campaign as part of a larger effort to demonstrate that "individuals can do more to stop pollution and litter, should do more themselves and criticize government and business less."[25]

Indeed, the Crying Indian commercial was part of a massive KAB publicity campaign to emphasize the role of individuals in fighting pollution. The advertisements repeatedly personalized the environmental crisis, suggesting that pollution emerged not from the decisions made by corporate and government elites, but rather from the "carelessness, indifference and bad habits" of individual Americans. One advertisement featured a freckle-faced girl with pigtails and inquisitive eyes asking her father: "Daddy, what did you do in the war against pollution?" (fig. 5.3). In another ad, KAB claimed that "kids" and "mommies" were equally to blame as "businessmen" and "vice presidents" for causing pollution. Blurring the line between the public and private spheres, these advertisements placed pollution squarely within the family domain. By personalizing the question of responsibility, the imagery reinforced the sense of collective guilt that audiences were meant to feel as they witnessed that single tear rolling—on television, across billboards, and in myriad newspapers and magazines—down the cheek of Iron Eyes Cody.[26]

Just as a number of Earth Day organizers rejected Pogo's statement, some leading environmentalists began to challenge the visual politics of KAB. When the Crying Indian spot debuted in 1971, KAB enjoyed the support of mainstream environmental groups, including the National Audubon Society, the Sierra Club, and the Wilderness Society. But these organizations all resigned from its advisory council by the mid-1970s. They objected to KAB for two reasons: its troubling political agenda and its penchant for visual obfuscation.

KAB clashed with these groups over an important environmental debate of the 1970s: an effort to pass "bottle bills," legislation that would require soft drink and beer producers to sell their beverages in reusable containers as they had done until quite recently. Indeed, the flip-top can and the disposable bottle were relatively new arrivals on the beverage scene. According to one study from 1976: "The throwaway container, which represented less than 10 percent of the soft drink market as recently as 1965, now represents nearly 70 percent of that market." The shift to the throwaway was responsible in part for the rising levels of litter that KAB publicized, but also, as environmentalists emphasized, for the mining of vast quantities of natural resources, the production of various kinds of pollution, and the generation of tremendous amounts of solid waste. The KAB leadership, composed of major corporations in the beverage and container industries, lined up

Daddy, what did you do in the war against pollution?

Of course you can always try to change the subject.

But one answer you can't give is that you weren't in it. Because in this war, there are no 4F's and no conscientious objectors. No deferments for married men or teen-agers. And no exemptions for women.

So like it or not, we're all in this one. But as the war heats up, millions of us stay coolly uninvolved. We have lots of alibis:

What can one person do?

It's up to "them" to do something about pollution — not me.

Besides, average people don't pollute. It's the corporations, institutions and municipalities.

The fact is that companies and governments are made up of people. It's people who make decisions and do things that foul up our water, land and air. And that goes for businessmen, government officials, housewives or homeowners.

What can one person do for the cause? Lots of things — maybe more than you think. Like cleaning your spark plugs every 1000 miles, using detergents in the recommended amounts, by upgrading incinerators to reduce smoke emissions, by proposing and supporting better waste treatment plants in your town. Yes, and throwing litter in a basket instead of in the street.

Above all, let's stop shifting the blame. People start pollution. People can stop it. When enough Americans realize this we'll have a fighting chance in the war against pollution.

People start pollution. People can stop it.

FIGURE 5.3. "Daddy, what did you do in the war against pollution?" Advertising Council / Keep America Beautiful advertisement, 1971. Courtesy of Ad Council Archives, University of Illinois, record series 13/2/207.

against the bottle bills, going so far in one case as to label supporters of such legislation as "Communists."[27]

KAB's opposition to bottle bills led many environmental groups to sever their ties with the organization and rebuke its advertising campaigns. A leader of Environmental Action, the group that helped sponsor the first Earth Day, explained how the single-minded focus on litter obscured larger environmental problems. "In reality, there is no *ecological* difference if an item is discarded in the street or if it is disposed of 'properly' in an open dump," he argued. "In either case it will never biodegrade, never be reused, and never cease to be an eyesore. In either case, the effort and energy which went into making the bottle or wrapper is wasted, and the raw materials used have been forever removed from the earth's limited supply." Likewise, a leader of the National Audubon Society dismissed Iron Eyes Cody as "the flip-top American Indian."[28]

But KAB continued to promulgate the image of the Ecological Indian. By the mid-1970s, an Ad Council official noted that "TV stations have continually asked for replacement films" of the 1971 commercial "because they have literally worn out the originals from the constant showings." In 1975 the Ad Council released a new television spot that featured Iron Eyes Cody, still outfitted in buckskins, riding a horse across the land to promote KAB. Although he still shed a tear, this new commercial also depicted scenes of environmental improvement and claimed that "some Americans today" display the same "simple reverence" for nature as the Native Americans did long ago. The Ecological Indian cast judgment but also provided reassurance. Having ruined the environment, modern Americans could now, KAB promised, experience a rebirth of native wisdom and even, another advertisement suggested, learn to chant an ancient prayer: "'Oh great spirit . . . make me walk in beauty! Make my heart respect all you have made.'"[29]

The Crying Indian functioned as a kind of secular jeremiad, his tear a liquid sermon on the nation's sinful disregard for the environment. His tear elicited feelings of guilt not from minority communities coping with the hazards of lead paint or farmworkers whose jobs required them to use pesticides, but rather from middle- and upper-class white Americans who worried about their complicity in the environmental crisis and romanticized the pristine past of the American Indian. Throughout the 1970s his face appeared frequently on television and in the print media, a constant reminder of the failure of Americans to accommodate themselves to the land. While many activists would question his message, the Crying Indian became for millions of Americans the quintessential symbol of environmentalism. Like Pogo, KAB presented ecological issues in moral, not structural, terms. The answer to pollution had nothing to do with power or economics or public policy; it was simply a matter of how individuals felt toward nature and how they acted in their daily lives. The Crying Indian reinforced the idea of governing at a distance to depict environmentalism as a moralistic cleanup crusade.

SIX
—

The Recycling Logo
and the Aesthetics of
Environmental Hope

Writing in *Science* magazine near the end of 1970, Leo
Marx, the American studies scholar and author of the influential *The Machine in the Garden* (1964), astutely recognized
the problems with mainstream framings of the environmental crisis. Marx began by discussing the visual media's
extensive portrayals of environmental problems during
the late 1960s, especially the widely circulated images of
the Santa Barbara oil spill: "The sight of lovely beaches
covered with crude oil, hundreds of dead and dying birds
trapped in the viscous stuff, had an incalculable effect
upon a mass audience." As Marx implied, the emotional
responses triggered by such imagery made environmentalism central to American public culture. Yet the media also
reduced the movement to "a bland, well-mannered, cleanup campaign." Too often, Marx argued, media outlets promulgated the idea "that ecological problems are in essence
technological, not political, and therefore easier to solve
than the problems of racism, war, or imperialism." In a
similar fashion, the reform measures of the environmental regulatory state struck Marx as "cosmetic." Although
certainly "worthwhile," these ameliorative policies failed
to confront the "expansionary ethos" of modern culture
and, he predicted, would prove unable to overcome "the
deeply entrenched, institutionalized character of the collective behavior" that caused environmental problems.[1]

Marx's insights apply to the larger paradox of environmental politics during this period: even as the state expanded its regulatory authority, the visual media urged Americans to solve the environmental crisis through changes in their individual behavior. These seemingly contradictory themes, though, worked together to bequeath a problematic legacy to popular environmentalism: one that successfully lowered lead levels in the ambient environment but did not strive to protect inner-city children from the hazards of lead paint; one that banned DDT but did not confront the dangers pesticides posed to farmworkers; one that enacted some changes to corporate behavior and government policy, but that too often made environmental concern appear a matter of personal guilt and individual responsibility—of picking up litter and recycling bottles and cans—rather than a struggle to transform the expansionary ethos of consumer capitalism. The twinned discourses of universal vulnerability and individual responsibility filtered out subaltern perspectives, ignored power relations, and naturalized a particular conception of environmentalism as a moralistic cleanup campaign.

The apparent paradox of environmental politics during this period also shows how the history of popular environmentalism does not correlate neatly with most historical accounts of the 1970s. Much of this scholarship seeks to define the cultural politics of the era around specific turning points—around moments, for example, when the United States moved in a more liberal or conservative direction. Thus, both Philip Jenkins and Edward Berkowitz argue that 1974 marked a "great divide" that signaled the end of liberal reform measures and a turn toward "more limited domestic policy." Both suggest as well that after this time the decade became increasingly characterized by a focus on individual responsibility for a wide range of issues. "If the sixties was an era of government grants to fix social problems and regulatory laws to assure proper behavior," Berkowitz argues, "the seventies was a time in which people rediscovered the power of . . . individual responsibility and raised questions about the effectiveness of regulation to change behavior in a desired way."[2]

Rather than revealing absolute shifts—*from* liberal reform *to* individual responsibility—the environmental politics of this period demonstrate instead how these themes complemented one another. Moral appeals about individual responsibility reinforced the expanding power of the regulatory state—and vice versa. In both cases, pollution and other environmental issues were seen as constituting a "crisis," a visible, definable problem that could be "solved"—on the one hand, through changes in individual behavior; and on the other, as Marx indicated, through technical fixes initiated by the federal government. In both cases, environmentalism was portrayed as a movement devoted to a specific entity—the "environment"—and not a broad-based effort to bring about social justice. Issues of race, class, and power, therefore, had nothing to do with the quest for a cleaner environment. Neutral experts employed by the environmental policy apparatus, together with individuals engaged in voluntary action around the nation, would participate in a consensus-building effort to rid America of pollution.[3]

Out of all the images that Earth Day would bequeath to the environmental imagination, none would be more significant than the recycling logo, a symbol that may well be the most widely seen image in contemporary culture, encountered daily in a plethora of public spaces and on a tremendous array of containers and packages. The recycling logo attempts to hold the apparently contradictory trends of individual responsibility and environmental regulation perfectly in balance, and the story of its creation helps reveal the limits of popular environmentalism.

On Earth Day 1970, Gary Anderson, a student at the University of Southern California, participated in an environmental rally on campus. He remembers that there was "definitely something in the air . . . that was beginning to color everyone's image of the earth and its resources. Neither, people were beginning to realize, was infinite." Anderson soon learned of a contest sponsored by the Container Corporation of America (CCA), a leading manufacturer of cardboard boxes and other packaging products, to design a symbol for recycling. The CCA appealed to the idea of futurity by opening the contest only to high school and college students, because, the company explained, "as inheritors of the earth, they should have their say." Anderson's winning entry—three chasing arrows that form an endless loop—brought together ecological ideas and countercultural aesthetics to convey the hope that the ritual of recycling could solve the environmental crisis (fig. 6.1).[4]

Anderson's visual connection to the counterculture came through the crucial influence of the Dutch artist M. C. Escher, whose work circulated widely during this period. In particular, Anderson's design for the recycling logo used a visual motif popularized by Escher: the Möbius strip. As one scholar explains, the Möbius strip features a one-sided surface made "by joining the two ends of a strip of paper after giving one end a 180-degree twist. An ant could crawl from any point on the surface to any other point without ever crossing over an edge."

FIGURE 6.1. Recycling Logo prototype by Gary Anderson, 1970. Courtesy of Gary Anderson.

After learning about the Möbius strip from a mathematician, Escher employed the motif in three works, including, most famously, his 1963 piece titled *Möbius Strip II*, which portrays "an endless procession of ants treading the looped surface of a Möbius strip—a finite number of ants on a twisted ladder that doesn't go anywhere yet never ends" (fig. 6.2). Escher's interest in the Möbius strip reflected his ongoing fascination with the tensions between the finite and the infinite; he often depicted patterns of interlocking shapes and figures to represent the concept of infinity within a finite pictorial space.[5]

While Escher's work had long been admired by scientists and mathematicians, in the late 1960s it began to acquire a much larger fan base among the counterculture and the environmental movement. "Ecologists took him up and through ecology the counter-culture found him and lavished smothering love on him, using his prints as record covers, day-glo posters, T-shirts, [and] wall paper," *Arts Magazine* explained.

FIGURE 6.2. M. C. Escher, *Möbius Strip II*, 1963. © 2013 The M. C. Escher Company—The Netherlands. All rights reserved. www.mcescher.com.

Embraced and promoted by the *Whole Earth Catalog*, Escher was also profiled by the then-hip music magazine *Rolling Stone* in a 1970 piece that accounted for the "sudden rush on Escher" by emphasizing "the close parallel of his vision to themes of contemporary 'psychedelic' art." In the months leading up to Earth Day, Escher's art also appeared in special environmental issues published by the liberal *Saturday Review* and the radical *Ramparts*. In both cases the magazines reproduced the images without any comment or explanation; they simply sprinkled Escher prints throughout the pages. Apparently, the editors believed, the pictures could speak for themselves: images of birds changing into fish, frogs changing into birds, and other such works conveyed ideas of ecological harmony and the wonders of nature.[6]

Gary Anderson, already a fan of Escher, looked to the Möbius strip for inspiration. "The figure was designed as a Möbius strip to symbolize continuity within a finite entity," he explained. "I used the [logo's] arrows to give directionality to the symbol. . . . I wanted to suggest both the dynamic (things are changing) and the static (it's a static equilibrium, a permanent kind of thing)." Like Escher, Anderson sought to visualize infinity within a closed cycle, to evoke unending temporality within a finite space. His design twisted the three chasing arrows "so that if they were joined in a continuous ribbon, they would form a Möbius strip."[7]

Anderson's recycling symbol, which later became ubiquitous worldwide, helped citizens reimagine their relationship to the environment. The logo presented a new aesthetic of environmental hope, and empowered individuals to feel that they could play a vital role in ensuring sustainability. As the long time frames of environmental fear became normalized, the recycling symbol presented a reassuring vision of a constantly regenerating future. The ravenous use of resources could continue apace as long as people remembered to close the recycling loop and create, as the logo promised, a sense of ecological equilibrium, a permanent balance with the natural world. Rather than acting as evil defilers of nature, Americans could participate in a new form of virtuous consumption and find redemption through recycling.[8]

The judging for the recycling logo contest took place as part of the International Design Conference at Aspen (IDCA), an annual event cofounded by CCA president Walter Paepcke that sought to get "designers and executives together for their mutual benefit." For the IDCA board, whose members included Bauhaus artists and corporate patrons of modernism, "design was a problem-solving activity in the service of industry." However, as the art historian Alice Twemlow explains, the

1970 conference, held just a few months after the first Earth Day, led to "a collision between two very different conceptions of design." Taking as its theme "Environment by Design," the conference pitted the design professionals against a large contingent of protesters—students, environmentalists, countercultural artists, and French intellectuals—who, Twemlow notes, "targeted the conference's lack of political engagement [and] its flimsy grasp of pressing environmental issues. . . . In their view, design was not the promulgation of good taste or the upholding of professional values; it had much larger social, and specifically environmental, repercussions for which designers must claim responsibility."[9]

Cliff Humphrey, the founder of Berkeley-based Ecology Action, a radical environmental group that had recently been profiled in the *New York Times Magazine*, explained the oppositional stance of the protesters in his IDCA speech. Humphrey mobilized survival discourse to emphasize the stakes of the environmental struggle and to politicize the concept of design: "What we are talking about, then," he argued, "is manifesting by design a survival gap—a survival gap between the people on this planet and the ability of the life support system to support these people." While addressing the audience, Humphrey used material objects and visual images of environmental crisis—including a mound of garbage and a photograph of the whole earth—to deplore consumer culture's wastefulness and register concern about the threats to the planet's life support systems. Humphrey called for radical change; he wanted to move beyond the reformist impulse, beyond the notion of merely regulating industry and cleaning up pollution. He wanted instead to limit the expansionary ethos of contemporary capitalism: "If an item is made to be wasted, to be dumped on a dump, then don't make it! You know, if our youth can say 'Hell, no!' to the draft, then I think a few of you have to learn to say 'Hell, no' to some salesmen and to some developers."[10]

For their part, the French intellectuals, led by the soon-to-be-influential postmodernist Jean Baudrillard, extended Humphrey's critique to question not only the precepts of professional design but also the popular discourse of environmental crisis. In a statement authored by Baudrillard, the French group found mainstream views of environmentalism to be nothing more than ideological mystification. "It is not by accident that all the Western governments have now launched . . . this new crusade, and try to mobilize people's conscience by shouting apocalypse," Baudrillard charged. "We see that ultimately the real issue is not the survival of the human species but the survival of political power." Baudrillard lambasted the emotional politics of popular environmentalism. In particular, he rejected Pogo's message of environmental guilt and the widespread effort "[to give] a guilty feeling to the collective consciousness. (We have the enemy and he is us.) . . . This blackmail towards apocalypse and toward a mythic enemy who is in us and all around tends to create a false interdependence among individuals. Nothing better than a touch of ecology and catastrophe to unite the social classes."[11]

These radical critiques of the IDCA provide a broader context in which to view the emergence of the recycling symbol. By selecting Anderson's design as the winner, the judges validated and helped enshrine the narrow vision of environmentalism challenged by Humphrey, Baudrillard, and other protesters. The closed loop and the sense of infinity and stability provided by the Möbius strip generate an image of sustainability, and suggest that recycling can provide a technical solution to the environmental crisis. The symbol implies that the three arrows continue cycling forever, eternally returning the same natural resources back into consumer products without any expense of energy or waste of materials. This vision of perpetual harmony works to perpetuate the same expansionary ethos that Leo Marx and others questioned in 1970, and to delimit the meanings of popular environmentalism.

During 1971, in stories related to the first anniversary of Earth Day, the message of individual responsibility became central to mainstream coverage of the environmental movement. In a special report titled "Happy Birthday Earth Day!," *Look* magazine profiled a group of young activists in Modesto, California, including Cliff Humphrey. Soon after the IDCA conference and the selection of the recycling logo, Humphrey had relocated to Modesto, where he helped start a new recycling program in the city. *Look* claimed that Humphrey and other activists wanted to "stop blaming industry" and stop pushing for new legislation. While Humphrey had continually sought to combine personal responsibility with radical action, he now appeared apolitical, at least according to *Look*. "Working to pass a law, no matter how vital, is no substitute for putting out a lighter garbage can," the magazine quoted him as saying. *Look* included photographs of Cub Scouts carrying empty bottles to a local recycling center and of "well-heeled and soft-spoken doctors' wives" meeting to discuss pollution. "Ecology jumps the generation gap," one caption asserted, echoing the sentiment voiced a year earlier, when many commentators portrayed Earth Day as a unifying moment in American culture. Looking back to the movement's inaugural moment, the mass media framed environmentalism as nothing more than a "cleanup crusade," an effort to prove, as the Crying Indian advertisement claimed, that people could stop pollution.[12]

Humphrey sought to provide a hopeful alternative vision of the environmental future. "We're way past the doomsday thing," he explained to *Look*. "We're starting to build something." Through its pragmatic effort to foster environmental hope, the Modesto recycling project typified the do-it-yourself environmentalism associated with certain segments of the counterculture. Humphrey, who clutched a copy of the *Whole Earth Catalog* when he spoke at the IDCA conference, wanted to put into practice the ideas of this countercultural bible by making Modesto into a model of sustainability. The local struggles of Humphrey and many other activists across the United States would eventually make it possible for recycling to go mainstream, to become an everyday practice of environmental citizenship. As we will see in part 3 of this book, by the 1980s and 1990s numerous cities and counties adopted curbside recycling programs, and Anderson's now-iconic logo would be stamped across bins and would also appear on bottles, cans, and other recyclable containers. While Humphrey and other activists critiqued the wasteful practices of consumer culture, the beverage and packaging industries appropriated their vision of environmental hope to present recycling as the ultimate expression of environmental citizenship.[13]

Recycling popularized the discourse of individual responsibility and deflected attention from industry practices. "Recycling constitutes a profoundly individuated response to the problem of waste," the historian Ted Steinberg argues. "It works to the advantage of industry by pushing the costs of business onto the public to bear. It also shields industry from any restrictions on its methods of production." Indeed, as one environmental activist noted in the 1970s, recycling presented "a frightening potential for institutionalizing waste generation." While the recycling logo continues to foster environmental hope, its Escher-inspired image of ecological equilibrium also obscures corporate practices that perpetuate waste and undermine sustainability.[14]

During the period surrounding the first Earth Day, a series of environmental icons—from the Santa Barbara oil spill and the whole earth to people wearing gas masks, Pogo, the Crying Indian, and, later, the recycling logo—helped bring environmentalism into American public culture. These images encouraged audiences to see themselves as potential victims of an escalating long-term crisis and to reimagine their political world by adopting new habits of citizenship. Whether the "cleanup campaign" was carried out by the government, individuals, or some combination of the two, the discursive focus on "pollution" made the environmental crisis appear as a visible threat to human bodies and nonhuman nature. The ecological lens and the long time frames of fear became more accepted in American public life, while the media tropes of universal vulnerability and responsibility marginalized radical and subaltern perspectives on the crisis.[15]

The specific environmental problems associated with dominant energy systems also remained relatively unseen during this period. Following Santa Barbara, other petro-disasters—such as a 1970 spill in Tampa Bay and a 1971 spill in San Francisco Bay—generated more images of oil-damaged wildlife. Media coverage of both the Tampa and San Francisco spills shared with Pogo, the Crying Indian, and the recycling logo the notion of governing at a distance, the idea that individuals could engage in personal acts to remedy the environmental crisis. While coverage of the Santa Barbara spill occasionally emphasized the role of volunteers, the trope of individual agency—of rescuers saving helpless, oil-soaked creatures—became central to popular depictions of these later spills.

Life's article about the San Francisco spill featured a hopeful narrative of "longhairs and hardhats" joining together to rescue oil-soaked birds and cover the beaches with protective straw. "The long-haired and bearded young were . . . all over the beaches," *Life* noted, "but they had no monopoly of concern. Standard Oil of California turned out 700 of its employees, including the hard-hatted Chevron worker above, to help organize the beaches and provide supplies. . . . Whole families spent hours wading into the water after foundering sea birds." Published almost a year after both the Earth Day 1970 celebration and the infamous "Hard Hat Riots," in which construction workers had attacked antiwar protesters in New York City, *Life* emphasized the unifying potential of environmentalism, a consensual cause that could bring together hard hats and hippies. *Life*'s visual narrative once again privileged individual action to show citizens taking personal responsibility for the environmental future. Just as the recycling logo conveyed feelings of environmental hope, the images of individual volunteers bathing birds and shoveling straw indicated the affective dynamics of environmental action, the sense of emotional recovery generated by the cleanup crusade.[16]

Even as oil spills continued to be viewed as symbols of an escalating environmental crisis, the political and public responses to these events emphasized the short-term damage of oil washing ashore rather than the long-term systemic problems of increasing energy use and petro-dependency. Media images presented oil spills as shockingly sad but nevertheless generic signs of pollution. In this coverage, the accretive dangers of a socioeconomic system based upon high levels of oil consumption remained largely hidden from view.[17]

In 1973, with the onset of the OPEC oil embargo, popular fears of finitude became specifically linked to the energy crisis. Fuel shortages, together with raging debates over nuclear power and the push for renewable energy, played a crucial role in defining environmental politics for the remainder of the 1970s. Although energy issues became increasingly visible, environmentalists struggled against the limits of media temporality to try to convey the long-term meanings of multiple energy problems. While presidents and public service announcements would emphasize the idea of governing at a distance, of individuals doing their part to conserve energy, many environmental thinkers and activists would instead envision large-scale changes to dominant energy systems. Rather than internalizing environmental guilt, rather than accepting the limits of media imagery, they sought to promote environmental hope by picturing alternative energy paths.

PART TWO

Energy Crises and Emotional Politics

SEVEN

Gas Lines and Power Struggles

The lines stretched on and on. They snaked around gas stations, through parking lots, and down the streets, sometimes for miles. Drivers waited and waited, sometimes for hours. They needed the gas to get to work, to pick up their kids, to go grocery shopping. Many became increasingly impatient as they waited in these seemingly endless lines. On occasion tempers would flare and violence would break out.

As the 1973–74 Organization of Petroleum Exporting Countries (OPEC) oil embargo caused energy supplies to dwindle and prices to skyrocket, photographers and TV newscasters used two visual strategies to depict gas lines as signs of crisis. Aerial views provided a panoramic perspective on the immense numbers of cars queuing up, forming distinct geometries of frustration below—lines, curves, S-shaped patterns—an abundant accumulation of automobiles idly waiting for scarce supplies of fuel (fig. 7.1). These pictures revealed vast landscapes of asphalt dotted with automobiles, rendering human figures invisible or minuscule. In contrast, ground-level shots often focused on the faces of one or two drivers standing outside their cars. The lines reached behind them, extending far beyond the borders of the frame. The other drivers remained in their cars, so that the feelings of these individuals stood in for those of the entire group. Some looked listless, resigned to waiting. Others appeared enraged, with fists pumping in the air, staging a futile protest against the fuel shortage (fig. 7.2).[1]

FIGURE 7.1. Cars line up in two directions at a gas station in New York City. Photograph by Marty Lederhandler, 1973. Used by permission of AP Photo.

FIGURE 7.2. Angry driver waving fist during gasoline crisis, Chicago, 1974. Copyright Bettmann/Corbis.

Gas line imagery did more than merely record the experience of the energy crisis in a neutral, objective fashion, and did more than passively reflect the feelings of Americans waiting for fuel. As active rhetorical agents, the pictures also shaped popular perceptions of energy issues and helped define the meanings of citizenship in 1970s America. Gas line imagery framed the energy crisis as solely a question of supply. The newsworthiness of the images stemmed not from the history of transportation policy that made the automobile so dominant in American life, but rather from the inconvenience of gas lines and the shocking onset of fuel shortages. Depicting citizens primarily as consumers denied a resource they needed in daily life, the pictures visualized a limited model of citizenship: one that portrayed automobility as an unchangeable fact of American existence. Isolated from one another, encased in their vehicles, the drivers appeared as individual consumers, not as citizens capable of acting collectively to shape the energy future.[2]

In suggesting that the crisis was a temporary aberration, something that would end as soon as supplies returned to normal, the pictures also conformed to dominant patterns of media temporality. Rather than situating the crisis within a long-term perspective, rather than considering how the nation's recent environmental history and transportation policies had led to fossil-fuel dependency, the images fragmented time and detached the present moment from broader contexts to show a nation of drivers running on empty.

In a period marked by severe economic decline and increasing public cynicism, gas lines also contributed to the decade's broader emotional politics that emphasized anger and alienation over a collective sense of hope and possibility. Consider the legendary lines uttered by Howard Beale, the "mad prophet of the airwaves" in the 1976 Hollywood film *Network* (fig. 7.3). Beale, a news anchor played by Peter Finch, addresses his TV audience immediately following a report about OPEC price hikes. The frustrating experience of the gas lines echoes in his televised rant. After listing various social and economic problems, Beale exclaims: "I want you to get mad. I don't want you to protest. I don't want you to riot. I don't want you to write to your Congressman, because I wouldn't know what to tell you to write." Rather than taking political action, Beale urges his viewers to yell, "I'm as mad as hell, and I'm not going to take this anymore!" His fictional tirade provides psychological release but signifies the diminution of hope, the inability to imagine a collective vision of a better future. As one scholar explains, "*Network* seems to despair of . . . hope. For its main social motif

FIGURE 7.3. Peter Finch as Howard Beale: "I'm as mad as hell!" Frame capture from *Network*, directed by Sidney Lumet, 1976.

is an insistence on the atomization of the television audience and, by extension, of the American public in the regressively privatized, anti-collective 1970s."[3]

Network's portrait of atomized, alienated citizens resonated with many environmental activists. Rather than accept this sense of power-lessness and disillusionment, though, the activists sought to generate social hope, to use the visual media as an instrument to challenge corporate power and democratize public culture. Moving beyond frustration, they sought to fashion an alternative energy future and to overcome the debilitating emotional politics of 1970s America.

While gas lines became the first visual marker of energy crisis, Americans in the 1970s encountered many other images that made energy concerns central to popular perceptions of crisis, of living through a critical moment and confronting a widespread sense of economic decline and potential environmental catastrophe. Throughout the decade, presidents and public service advertisements berated and beseeched them to end their wasteful ways by making a concerted effort to conserve energy in everyday life. Protesters against nuclear power warned of possible meltdowns and other hazards of radioactivity, a message that became more widely accepted following the 1979 accident at Three Mile Island and the spectacular success of the Hollywood film *The China Syndrome*. Advocates of solar power hailed renewable energy as a way to solve the energy crisis and a means to decentralize and democratize the nation. Jimmy Carter installed solar panels on the

White House roof and delivered a series of televised addresses that presented energy issues as the moral equivalent of war.

During the time surrounding Earth Day 1970, pictures of the Santa Barbara oil spill and of people wearing gas masks had made pollution appear as an apocalyptic sign of environmental crisis, a threatening condition that turned everyone into a potential victim. The discursive focus on pollution helped bring environmentalism into American public culture, but tended to obscure the specific environmental problems caused by dominant energy systems. With the onset of the OPEC embargo, the growth of organized opposition to nuclear power, and the effort to promote renewable sources of power, energy issues became a defining feature of American environmental politics.

While gas masks had made Sarah, Lucy, and other Americans appear vulnerable to pollution, gas line imagery made consumers appear vulnerable to scarcity. Yet, as environmental activists insisted and the visual media sometimes indicated, the energy crisis extended well beyond the gas lines to raise broader questions about the nation's dominant energy and transportation systems: Should the crisis be understood simply as a temporary fuel shortage triggered by a momentary lack of supply? Or did it stem from an ongoing, escalating problem of excessive demand? Did nuclear power offer a path to energy independence? Or did it pose a grave danger, a technology that threatened porous ecological bodies? Throughout the remainder of the 1970s, the visual media figured prominently in larger struggles over the contested meanings of energy crisis. As environmentalists struggled to promote alternative energy ideas, they found themselves facing some familiar adversaries and challenging the limits of media representation.

The Ad Council responded to the OPEC embargo by launching an energy conservation campaign. Developed in conjunction with the Federal Energy Office, an executive agency created by Richard Nixon, the campaign repeatedly admonished audiences: "Don't be fuelish." TV commercials starring the actor George C. Scott and Miami Dolphins football coach Don Shula urged viewers to engage in energy-saving actions in daily life: lowering thermostats, turning off unused lights and appliances, and driving more slowly. Meanwhile, the cartoonist Jack Davis sketched a series of illustrations for billboards, magazines, and newspapers that showed individuals practicing various "ways not to be fuelish." Across these different media, the "Don't Be Fuelish" campaign promoted the idea of governing at a distance, of individual Americans adopting practices of self-regulation to demonstrate their citizenship values.[4]

For many environmentalists, the new Ad Council campaign seemed like the Crying Indian all over again. Even as President Nixon instructed Americans, "in the short run," to "use less energy . . . less heat, less electricity, less gasoline," his long-term plan—known as Project Independence—envisioned the nation producing "new sources of energy": more oil drilling, more coal mining, more nuclear power plants. At the same time that Nixon and the visual media promoted personal solutions to the energy crisis, government and corporate policies pushed for the extensive development of what environmentalists considered to be destructive, nonrenewable sources of power. Like the Keep America Beautiful campaign, the catchphrase "Don't be fuelish" emphasized individual responsibility but seemed to let corporations and governments off the hook. While environmental activists certainly agreed that individuals should reduce their energy consumption, they believed that the Ad Council once again was engaging in ideological mystification. Its energy conservation messages deflected responsibility and obscured other possible solutions to the energy crisis.[5]

Following the Watergate scandal and Nixon's resignation, President Gerald Ford continued to frame the energy crisis as a question of supply, and used his 1975 State of the Union address to trumpet his new energy plan. "Within the next 10 years," Ford announced, "my program envisions: 200 major nuclear power plants; 250 major new coal mines; 150 major coal-fired power plants; 30 major new [oil] refineries; 20 major new synthetic fuel plants; [and] the drilling of many thousands of new oil wells." Soon after Ford's speech, the Ad Council released another set of TV commercials. Known as the "Little Boy" spots, these ads deployed a familiar environmentalist strategy: they depicted a young white child imploring "grown-ups to save America for his generation." Just as environmentalists had used images of children to represent futurity, these spots conveyed the "boy's emotional plea" by showing him at various historic landmarks, "wondering what America would be like if adults continued to waste fuel." In one ad, the boy stood near the gleaming torch of the Statue of Liberty and announced his fear of future scarcity. The camera then zoomed out to reveal a fuller view of the iconic statue, as the campaign's tagline appeared on the screen (fig. 7.4).

No tear was shed in these commercials, but the Ad Council once again sought to elicit feelings of guilt, to make viewers feel responsible for the nation's environmental problems. Despite the campaign's focus on energy rather than pollution and litter, the message remained the same: People started this crisis, and people could stop it.[6]

DON'T BE FUELISH
SAVE ENERGY

FIGURE 7.4. "Don't Be Fuelish." Frame capture from Advertising Council / Federal Energy Office television ad campaign, 1975.

In 1976, the Ad Council and Keep America Beautiful (KAB) brought back Iron Eyes Cody as the star of a new antilitter campaign. The Crying Indian's reappearance enraged many environmentalists, including Earth Day veteran Peter Harnik. Like other critics of the Crying Indian, Harnik emphasized the ecological problems of throwaway containers and rebuked KAB for its opposition to bottle bills. Summarizing the new KAB campaign, Harnik noted that the "the same sad Indian" would appear, but he would see people picking up litter and begin to smile. Yet "the real reason he will be smiling," Harnik sardonically concluded, "[is] because his industrial bosses can continue to use throwaway bottles and cans."[7]

As the Crying Indian reappeared while KAB continued to fight against bottle bills, as the Ad Council told Americans not to be "fuelish" while Nixon and Ford furthered the national dependence on nonrenewable energies, environmentalists began to wonder: Maybe they could challenge the power of the Ad Council by making their own public service announcements (PSAs). Maybe they could move beyond the relentless focus on individual action to present alternative views of environmental problems. Maybe they could democratize the airwaves, galvanize viewing audiences, and, in turn, transcend the debilitating emotional politics of the era.

Environmental activists wanted to produce PSAs that would address a wide range of energy issues. First, they wanted to publicize the dangers of nuclear power. When introduced in the 1950s, commercial nuclear power had been praised in exorbitant terms—Atomic Energy Commission Chairman Lewis Strauss famously declared that it would lead to "electrical energy too cheap to meter"—but the industry grew slowly, with only a few reactors contributing to the electrical grid by the early 1960s. In the middle of that decade, though, the industry experienced a sudden surge in reactor construction and new plant orders. According to its proponents, nuclear power offered the United States a safe, sustainable form of energy independence. The industry's planned expansion, however, led to rising opposition by local grassroots groups as well as national environmental organizations. Emphasizing the shared threats to human bodies and nonhuman nature, the antireactor movement expressed its concern through the ecological lens that had been popularized by Rachel Carson and the first Earth Day. Indeed, the names of the local groups—including the Clamshell Alliance, the Abalone Alliance, and the Red Clover Alliance—all indicated their concern with the wider ecologies of radioactive risk. The antireactor movement also drew on the trope of the vulnerable child to depict the dangers to futurity. Following the visual tradition established by SANE and other antinuclear groups, one popular protest poster from the 1970s featured an image of a mother and toddler, along with the caption: "What do you do in case of a nuclear accident? Kiss your children goodbye."[8]

Environmentalists wanted to use PSAs not only to visualize nuclear fear, but also to circulate hopeful visions of the future. They embraced what the alternative energy expert Amory Lovins described as the "soft path." According to Lovins, US energy policy was following the "hard path," marked by the "rapid expansion of centralized high technologies to increase supplies of energy." The hard path's commitment to nonrenewable practices emphasized the "short term" at the expense of "long-term sustainability." Rather than relying upon the hard path of nuclear energy and fossil fuels, Lovins instead promoted a "soft path" based on "a prompt and serious commitment to efficient use of energy" and the "rapid development of renewable energy sources."[9]

While the Ad Council exhorted Americans to reduce their personal energy consumption, environmental groups produced PSAs that challenged dominant portrayals of the energy crisis and sought to imagine an alternative energy future. One PSA directly responded to the Ad Council's "Don't Be Fuelish" campaign. The spot began by announcing, "It's not your fault," and then explained how the crisis was caused not by individual Americans but rather by "long-term mismanagement on the part of the energy industry and government." PSAs that opposed nuclear power used fear to condemn this technology: the spots featured the tag line, "Nuclear power is a terrible way to go," ominously inscribed on a tombstone. Other PSAs simultaneously protested nuclear power and promoted renewable energy sources. These spots moved beyond mainstream depictions of the energy crisis—which focused on insufficient supply in the wake of the OPEC embargo—to consider the nation's overreliance on fossil fuels and the need to develop alternative energies.[10]

Although environmental groups hoped that PSAs would enable them to reframe the energy crisis, they were often denied airtime and were unable to broadcast their messages. The problem, they argued, was that the Ad Council possessed a "virtual monopoly of public service time" and exploited its power to silence oppositional voices. Not only did the Ad Council dominate the public service field, they charged, but many broadcasters refused to run PSAs that departed from its ostensibly apolitical approach. "By jamming the media with its views," one commentator explained, "the Ad Council . . . manages to prevent any real critical views from receiving free and widespread dissemination."[11]

To challenge the Ad Council's dominance, the Media Access Project, a public interest law firm representing sixty-five citizen organizations, filed a petition with the Federal Communications Commission (FCC) to "open up public service announcements to a myriad of local and national groups with fresh, creative, alternative approaches to the issues confronting Americans." While the petitioners hailed from a diverse array of progressive organizations—including the National Gay Task Force, the National Council of La Raza, and the National Organization for Women—environmentalism encompassed the largest contingent of signatories. Indeed, out of the sixty-five groups, at least fifteen primarily focused on environmental issues, including mainstream nationals like the Sierra Club, more radical nationals like Friends of the Earth and Environmental Action, and a plethora of local organizations.[12]

The petition repeatedly cited the Crying Indian as a prime example of the Ad Council's problematic control over the PSA field. "These anti-littering messages . . . divert (and seem designed to divert) public attention from the problems of industry's contribution to pollution," the petition claimed. Like other Ad Council campaigns, the Keep America Beautiful spots offered a "one-dimensional approach which suggests that the solution to some major problems confronting Americans lies in individual action."[13]

This petition offers a revealing glimpse at how environmental activists viewed the nexus of media, citizenship, and public life during the 1970s, and indicates how they sought to overcome the dispiriting vision expressed by *Network* and other cultural texts. "It is a fact of American public life in the 1970's that the public is disillusioned, disenchanted and alienated from the political, social, economic and cultural forces shaping the lives of the American people," the petition argued. "Such cynicism breeds the alienation and apathy which is the very antithesis of wide participation in public life." Throughout the 1970s, commentators from across the political spectrum noted the growing sense of apathy, cynicism, and frustration in American life, the widespread feelings of withdrawal from public culture, and the declining trust in leaders and the government. Environmentalists believed that new PSA policies would enable them to disseminate their empowering ideas to broader publics and thus counter the alienation and apathy of the era.[14]

Not surprisingly, the Ad Council rejected the petition's claims. While activist groups believed that new PSA policies would promote media democracy, the Ad Council argued that their proposals would instead lead to media anarchy. The new rules, the Ad Council claimed, would violate the notion of "public *service* announcements" by opening up the airwaves to "public *debating* announcements." By requiring TV stations to "provide time for a plethora of advocacy positions," the airwaves would be transformed into "a veritable tower of Babel from a myriad of interested parties."[15]

In their call for new PSA rules, environmentalists and other petitioners directly challenged the Ad Council's definition of political speech. While the Ad Council denigrated antinuclear and pro-solar spots as misguided attempts to politicize this supposedly neutral form of address, the petitioners insisted that the Crying Indian and other campaigns emphasizing individual responsibility also projected a particular political vision. By obscuring alternative explanations of causes and potential solutions, these PSAs constricted the idea of environmental citizenship to be nothing more than individuals doing their part—whether it be, in the case of Keep America Beautiful, picking up litter or, in the case of energy conservation, trying not to be "fuelish."

The FCC denied the petition, and PSA policy remained essentially the same. While environmentalists had hoped that the PSA could become "a highly valuable resource for citizen groups," this mode of communication remained closed off and controlled by the Ad Council. Although they failed to democratize the airwaves, although they remained haunted by the Crying Indian, environmentalists engaged in other power struggles that encouraged the popular media to widen the frame of energy crisis, and to move beyond the iconic gas lines to cover such issues as the nuclear threat to human health and the promising allure of the sun.[16]

The next four chapters will consider the late 1970s as a pivotal environmental moment in which energy crises became central to American public culture. By 1979 it would appear that the news media, Hollywood films, and the televised speeches of Carter all worked to popularize environmental concern and legitimate soft-path critiques of dominant energy systems. The intense media attention given to Three Mile Island, the blockbuster success of *The China Syndrome*, and a series of presidential pronouncements all suggested that radical energy ideas had become mainstream, and that the visual media had proved conducive to disseminating their far-reaching proposals to alter the nation's energy future.

In response to these images, conservative pundits would develop a critique of popular environmentalism that has become a recurring feature of American public culture. Conservatives claimed that media imagery manipulated audience emotion and duped the public into accepting environmentalist fear-mongering. Invoking the familiar dualisms of reason versus emotion, fact versus feeling, science versus spectacle, they would point to *The China Syndrome* and news coverage of the Three Mile Island accident as sensationalistic imagery that falsely discredited nuclear power. Ever since this period, conservatives have dismissed other examples of popular environmentalism in similar terms.

Yet even as the visual media helped legitimate certain environmental concerns, these same channels of communication also worked to narrow popular framings of environmentalism. In their antireactor campaigns, environmental activists emphasized the multiple temporalities of danger. They worried that low-level releases of radioactivity caused by the everyday functioning of reactors threatened the long-term well-being of the local ecology and the bodily health of nearby residents, especially children. They also warned of the extensive time scale of toxicity represented by nuclear waste, and questioned whether these radioactive by-products could be stored safely for the duration of their multimillennial half-lives. While the movement registered its concerns in multiple temporal frames, the sudden potential for catastrophic accidents became, especially in the aftermath of Three Mile Island, the iconic and most widely circulated representation of nuclear danger. Sudden violence trumped slow violence, thus obscuring the long-term risks associated with nuclear power production. Moreover, in promoting the soft path, some environmentalists sought to foster a collective vision of environmental hope. Joining forces with labor, civil rights, and other social movement leaders, these activists presented a cross-class vision of the alternative energy movement to counter the emerging media stereotype of environmentalism as an elitist anti-jobs campaign. Yet their efforts to emphasize the collective meanings of alternative energy failed to overcome the dominant media frames of individual hope and personal responsibility for the environmental future.

The emotional politics of the 1970s—including the climate of cynicism and the growing distrust of experts—intersected with popular environmentalism in shifting and surprising ways. The emotionally charged struggle to prevent further nuclear plant construction eventually aligned with dominant patterns of media temporality. The sense of urgency represented by *The China Syndrome* and Three Mile Island fused with the defensive, reactive dimension of environmentalism to mobilize popular anxiety toward nuclear power. Even as this coverage concentrated public attention on meltdown fears, media images rendered invisible the slow violence of dominant energy systems and slighted the environmentalist effort to envision a hopeful, sustainable energy future. While media profiles of solar inventors increased popular interest in renewable energy, this coverage, like the Ad Council spots and presidential speeches, emphasized the individualist dimension of environmental concern through personal stories and isolated glimpses of technical innovation—not on the prospect of a new energy path emerging out of the crisis. During this crucial environmental moment, the visual media conveyed the dangers of radiation to a mass public, but reaffirmed popular conceptions of environmentalism as a cause that focused on individual moral choices rather than on broader structural solutions.

EIGHT

Nuclear Meltdown I:
The China Syndrome

It was, up until that time, one of the most heavily promoted movies in Hollywood history. Three weeks before
The China Syndrome (1979) opened, Columbia Pictures
launched what it described as "the 'ultimate tease campaign' to hype" the film. A key component of the advertising blitz played off the mystery and uncertainty surrounding the film's title. As the company's vice president
for publicity explained, "The minute someone says, 'What
does the title mean?' they're halfway to buying a ticket."
Mounting a massive "tease-by-television campaign," Columbia spent close to three million dollars on commercials to heighten audience curiosity. Promotional materials
showed "only a fiery ball that could be the birth or death
throes of some giant star but is actually meant to simulate
the inside of a nuclear reactor." This mysterious image appeared along with pictures of the film's three stars—Jane
Fonda, Jack Lemmon, and Michael Douglas (fig. 8.1). The
film trailer announced: "'The China Syndrome.' Today
only a handful of people know what it really means, and
they are scared. Soon you will know."[1]

Even before opening night, *The China Syndrome* sparked
controversy. "The nuclear power industry's reaction . . .
was so swift that it preceded the release of the movie by
several months," the *Los Angeles Times* observed. Industry trade groups blanketed film reviewers, newspaper
editors, and TV broadcasters with pronuclear materials to condemn *The China Syndrome* for rendering "an

FIGURE 8.1. Poster for *The China Syndrome*, 1979. © 1979 Columbia Pictures Industries Inc. All rights reserved. Courtesy of Columbia Pictures.

unconscionable public disservice by using phony theatrics to frighten Americans away from a desperately needed energy source." An industry trade magazine warned: "Just in case you don't have enough grief, this 'contemporary thriller' will open in your service area March 16." The magazine then skewered the leftist politics of the filmmakers and actors: "As for the slant, consider the proclivities of the principals and draw your own conclusions—co-stars Jane Fonda and Jack Lemmon are anti-nuclear activists; executive producer Bruce Gilbert worked in support of Daniel Ellsberg and Tony Russo in the Pentagon Papers trial; Mike Gray, who wrote the original screenplay, includes among his credits 'The Murder of Fred Hampton,' a documentary on the Black Panthers." According to the nuclear industry, radical activists had teamed up to discredit nuclear power, seeking to use Hollywood film as a vehicle to propagandize their New Left–inflected and emotion-laden politics. The industry's fear of the film would lead one antinuclear scientist to respond: "If [pronuclear groups] believe that the production of one fictional feature film is going to harm their industry, then there really is something wrong with it. I mean, I couldn't think of another industry that would go into convulsions like this over one movie."[2]

Before March 16, only a handful of people may have known what the title meant, but record-setting audiences would learn that weekend, as *The China Syndrome* generated "the biggest non-holiday weekend gross in company annals." As moviegoers flocked to theaters, most reviewers praised the film. Vincent Canby of the *New York Times*, for example, hailed it as a "smashingly effective, very stylish suspense melodrama." Likewise, Richard Schickel of *Time* described it as "a superbly successful, expertly crafted, entirely riveting entertainment."[3]

While most critics agreed that the film provided riveting entertainment, *The China Syndrome*'s claims on the public imagination hinged on larger debates over its verisimilitude and underlying politics, debates that often boiled down to the following question: Should audiences accept the film's frightening depiction of nuclear power? Two days after it opened, the *New York Times* invited six nuclear experts to evaluate "this fictional but realistic film." A Westinghouse executive rebuked it for being marred by "technical flaws" and for developing "an overall character assassination of an entire industry." The movie, he argued, unfairly maligned "the utility chairman and the plant foreman . . . as morally corrupt and insensitive to their responsibilities to society." In contrast, a leader of the Union of Concerned Scientists, an antinuclear group based at the Massachussetts Institute of Technology (MIT), emphasized the film's basis in reality, arguing that it presents "a composite of real events and provides a scenario that is completely plausible." *The China Syndrome*, he concluded, was "a major corrective to the myth that was drilled into us as children, that nuclear energy is a beautiful, endless, cheap source of electricity. The movie shows how that dream has been perverted by companies . . . and how susceptible the program is to human error and industrial malfeasance."[4]

While questions about the blurring of fact and fiction shaped the initial reception of *The China Syndrome*, the notorious accident at a nuclear power plant on Three Mile Island in Pennsylvania—on March 28, only twelve days after the movie's release—made the intersection of film and reality even more salient and politically charged. The remarkable, almost simultaneous coincidence of disaster flick and real-life crisis marked an unprecedented occurrence in film history that yielded tremendous publicity for the movie. As one producer put it: "It's like making a movie about Pearl Harbor two days before the sneak attack." Suddenly, after Three Mile Island, the film seemed prophetic: the *New York Times* would run a story titled "When Nuclear Crisis Imitates a Film," while *Variety* would describe the crisis in Pennsylvania as a "powerful trailer" for *The China Syndrome*.[5]

As *The China Syndrome* continued to play in movie theaters, TV audiences watched nightly news broadcasts about Three Mile Island, reports that emphasized the possibility of a core meltdown in the plant. News broadcasters and elected officials directly referenced the film, as they began to wonder whether the crisis would turn into a catastrophe, whether Three Mile Island would become a "China syndrome" situation. In turn, the accident enhanced the oppositional status and truth claims of the movie and thus allowed it do to far more cultural and political work than it could have ever done without the real crisis. Together, *The China Syndrome* and Three Mile Island would lead the mainstream media to modify the dominant frames previously applied to the antireactor movement, to make the protesters appear more legitimate in articulating their fears about nuclear power.

With its focus on crisis and the threat of immediate catastrophe, media temporality advanced the emotional politics of the antinuclear cause. Through its coverage of Three Mile Island, the media revealed the risk to vulnerable ecological bodies and concentrated public attention on the unfolding drama of the accident. Television imparted a sense of urgency that emphasized the potential explosiveness of the situation. News broadcasts made spectators feel like witnesses to a crisis that could at any moment turn into a deadly catastrophe. Soon after the crisis ended, photographs on the covers of *Time, Newsweek, Life*, and other magazines presented the Three Mile Island cooling towers—hauntingly hovering over the landscape—as frightening reminders of the accident. The towers became iconic emblems that continue to warn of possible catastrophes at nuclear plants.

Yet even as the visual media reinforced one dimension of the antinuclear critique, it simultaneously ignored and marginalized other concerns of this movement. The narrative of *The China Syndrome*, the TV coverage of Three Mile Island, and the iconic images of cooling towers all focused on the nuclear reactor as the sole locus of danger, as the exclusive site of radioactive risk. The compressed timescale of the media foreshortened fear and ignored the broader spatial and temporal geographies of nuclear danger: from the mining of uranium that threatened the health of workers and Navajo communities in the Southwest to the storage of nuclear waste products, radioactive materials that persist for thousands of years. This narrowing of focus, this emphasis on the potential catastrophe of a China syndrome, galvanized public fear but fragmented public understanding, detached the possibility of meltdown from other nuclear risks and thereby constricted the temporal and spatial horizons of environmental citizenship. Indeed, despite its radical intentions, *The China Syndrome* received some of its harshest criticism from leftist reviewers, who claimed that the film failed to foster activist feelings or to imagine citizens playing any role in shaping the energy future.

Beginning on March 28, though, the movie suddenly seemed eerily prophetic. The cinematic text, refracted through the events at Three Mile Island, took on new meaning, moving beyond the limits of fictional film to express the reality of nuclear risk. The spectacle of mass culture merged with the spectacle of media coverage to make the movie's premonition of nuclear meltdown into an oppositional vision. As the crisis at Three Mile Island seemed to imitate the film, *The China Syndrome* became a surprising tool of protest, a movie that circulated beyond the screen and outside the theater to popularize and advance the antireactor movement.

Even before the film's shocking convergence with Three Mile Island, the entire story of the making of *The China Syndrome* involved a series of overlaps between fiction and reality. The radical documentary filmmaker Mike Gray developed the initial idea for the film in the early 1970s. As he drafted the script, Gray researched the dangers of nuclear power and decided to incorporate details from two recent accidents—a stuck needle on a gauge at an Illinois plant and dangerously low water levels that "almost uncovered the core" of an Alabama reactor—into a fictional story that emphasized the possibility of a nuclear meltdown. The more Gray learned about nuclear power, the more convinced he became that a catastrophic accident might occur soon. "I was running out of time," Gray explained. "I felt sure there would be a really bad nuclear accident. I had to beat that accident."[6]

As he worked on the script, Gray met with Henry Kendall, a Nobel Prize–winning physicist and a leader of the Union of Concerned Scientists. From Kendall he learned about the frightening conclusions of the Brookhaven Report—an Atomic Energy Commission study completed in the 1950s, revised in the mid-1960s but then suppressed until environmentalists filed a Freedom of Information Act request in 1973. The Brookhaven Report pondered the potential consequences of a worst-case scenario at a nuclear plant—a core meltdown that burned through the containment vessel and released dangerous amounts of radioactivity. Such an accident, the report concluded, would kill forty-five thousand people, injure another hundred thousand, and contaminate a massive area, "equal to that of the state of Pennsylvania." Gray lifted this line from the Brookhaven Report and revised it for his script. Given Three Mile Island's location in, of all places, the state of Pennsylvania, this line would be considered the most eerie moment in *The China Syndrome*.[7]

Despite his documentary experience, Gray decided to infuse his antinuclear concern into a fictional film. "I had done documentaries," he explained to an interviewer, "and I found a limit in them. The really essential ones you can't get on TV. The advertisers control the news." Gray's critique of TV echoed the claims made by environmental groups in their FCC petition about PSA policy. In contrast, he hoped that Hollywood would allow him to express dissenting views and politicize mass audiences.[8]

Gray's script eventually reached Michael Douglas, who decided to produce the film. Douglas had recently coproduced *One Flew over the Cuckoo's Nest* (1975), which had garnered five Oscars, but found it difficult to obtain studio support for the nuclear power film. According to Douglas, Gray's lack of experience in fictional film, together with the "controversial subject matter" that did not spell easy "commercial success," made Hollywood studios wary of committing to the project.[9]

Douglas then learned that Columbia Pictures had been negotiating with Jane Fonda about another nuclear-related film project to be produced by IPC, her film production company. Fonda had formed IPC in 1973 with Bruce Gilbert "to make the kind of movies which

Hollywood should be making but wasn't." Gilbert and Fonda had met through their mutual involvement in the Indochina Peace Campaign, an antiwar group whose acronym they adopted as the name of their film company. Gilbert believed in the radical potential of mass culture, and thought that Hollywood movies could "communicate progressive ideas to a mass audience." Such films, though, needed to appeal to general audiences; they should not be documentaries or avant-garde productions consigned to arthouse theaters. Together, Fonda and Gilbert developed a strategy to politicize viewers, a model they would apply to *The China Syndrome*. This approach, Fonda's biographer explains, would "start with an ordinary individual solidly wedded to society's traditional values and beliefs. Place that character in a story whereby he, or she, as a result of the logical flow of appropriate events, undergoes a deep emotional experience and become[s] . . . awakened to the reality he or she never realized existed before. . . . In that way, the character will gain the audience's sympathy." This narrative strategy, Fonda and Gilbert believed, would politicize audiences, thus turning Hollywood into an instrument of progressive activism. Hoping to make a movie about nuclear power but stymied by problems with their own project, they agreed to collaborate with Douglas to produce *The China Syndrome*.[10]

Before starring in *The China Syndrome*, Jack Lemmon had also encountered limits in the documentary mode: he had worked on an antireactor documentary that was shelved following a lawsuit filed by California's largest utility company. Titled *Powers That Be* (1971) and produced by Don Widener, the documentary aired once on a Los Angeles station. NBC then planned to broadcast *Powers That Be* nationally, but Pacific Gas and Electric Company (PG&E) sued Widener, charging that he had falsely dubbed in an interview segment. Even though PG&E lost the suit (and, in fact, Widener successfully countersued for libel), the long legal proceedings prevented the film from ever being aired again. Widener described the saga as a blatant example of industry intimidation. "There is an ongoing effort to suppress all antinuclear media coverage," he claimed. "Anyone trying to produce nuclear films hears from the industry."[11]

If television and public service announcements seemed unduly restricting, if they seemed to block democratic debate and stifle dissenting views, then perhaps Hollywood film would offer a counterhegemonic possibility, and would allow nuclear power's critics a chance to circulate their message to broader audiences. That prospect motivated Gray to write the script, Gilbert and others to produce it, and Fonda and Lemmon to star in the film. Although Gray had planned to direct the film, he expressed "stylistic differences" with Fonda and eventually gave up this role. James Bridges, who had previously directed *The Paper Chase* (1973), took over and also significantly revised Gray's script. Among other changes, Bridges introduced a female lead to be played by Fonda: a TV reporter who becomes increasingly concerned about the risk of a reactor meltdown.[12]

The China Syndrome mobilized fear of nuclear power by using the narrative structure of the conspiracy thriller, a film genre that flourished in the 1970s and powerfully conveyed the emotional politics of the era. "In both mood and theme," the film historian David Cook explains, "the conspiracy film was a type of paranoid political thriller that placed the blame for American society's corruption on plotters pursuing secret agendas to control national life." The Watergate scandal, the military defeat in Vietnam, and the decade's severe economic recession all contributed to growing disillusionment and distrust of the government and other institutions. Films like *The Parallax View* (1974), *The Conversation* (1974), and *All the President's Men* (1976) expressed popular feelings of alienation and cynicism by portraying an elite cabal that conceals the truth and conspires to implement its evil agenda. *The China Syndrome* not only depicted nuclear plant officials as scheming and murderous but also critiqued the mass media—specifically TV news—for its culpability in the growth of this industry, and for its failure to inform the public about the hazards of nuclear power. Throughout most of the film, network managers and nuclear executives collude to keep spectators from seeing evidence of radioactive risk.[13]

In the film, Fonda stars as Kimberly Wells, a newscaster in southern California. Given the sexist practices of the TV station, Kimberly is relegated to "happy talk," fluff pieces on such topics as singing telegrams and a tiger's birthday party at the zoo. Meanwhile, in the control room, the station managers—all of whom are men—talk about how her good looks and cheerful disposition have boosted their ratings. Although they believe she is compliant and will do whatever she is told, Kimberly desperately wants to break through the glass ceiling and start covering serious news stories.

The station managers then assign her to report about energy issues in California and assume she will not ask challenging questions. Along with cameraman Richard Adams, played by Michael Douglas, Kimberly visits a nuclear power plant. The public relations officer treats them to a slick, unabashedly positive presentation about the virtues of nuclear energy and the impeccable safety record of the plant. While Kimberly seems to accept the information uncritically, Richard—described in Bridges's script as "an intense young man with a beard," someone "who spent his formative years in the 60's" and who "always has a cause"— makes wisecracks and appears petulant in his dismissal of the PR man's shiny, happy portrait of nuclear power. As part of the tour, they enter the visitor's gallery overlooking the control room, where Jack Godell, played by Jack Lemmon, works as the shift supervisor. The guide informs Richard that he cannot use his camera there "for obvious security reasons." Suddenly, a turbine trip creates chaos in the control room. The needle on a gauge gets stuck, and panic grips the faces of Jack and other employees. Richard secretly films the entire episode (fig. 8.2).[14]

Kimberly and Richard return to the station, unsure of exactly what has happened, but certain that they have witnessed something alarming and newsworthy, something the public deserves to see. The camera in *The China Syndrome* appears as a truth-telling device, a machine that can render visible the hidden workings of nuclear reactors and warn audiences of the potential for a catastrophic accident. Yet the programming manager, after viewing the footage, refuses to run the tape and even orders that it be locked up in a vault. These scenes reinforced the view held by many environmental activists that the mainstream media

FIGURE 8.2. Michael Douglas and Jane Fonda in nuclear power plant. Frame capture from *The China Syndrome*, directed by James Bridges, 1979.

did not foster democracy but rather concealed truths from the public and stifled dissent.

Although Kimberly expresses mild disappointment at the decision, Richard is outraged. Indeed, he can hardly contain himself. In one heated exchange, he yells at the programming manager: "You're being pressured!"

"And you're being hysterical, Richard," the manager retorts.

"And you're being a chickenshit asshole!" Richard exclaims. Believing that the cameraman has crossed a line, Kimberly admonishes him and urges him to be more tactful.

Together with other key moments in the film, this conversation indicates how political feelings are negotiated and valued in *The China Syndrome*. Richard, the quintessential sixties radical, seems too hotheaded, too quick to condemn nuclear technology and challenge authority figures. Even though his critical views are ultimately validated by the end of the movie, his "hysterical" reactions appear juvenile, based solely on feeling and not on fact. In the early stages of *The China Syndrome*, the film questions Richard's emotional politics and presents his sixties radicalism as passé and outmoded. In contrast, the film legitimates and rewards the feelings of both Kimberly and Jack. As the plot develops, it makes their growing skepticism appear grounded in empirical reality, a logical outgrowth of the disturbing evidence they uncover.

In an interview conducted years after *The China Syndrome*'s release, Fonda described the nexus of emotions and dissent that operate in the film. Together with Bruce Gilbert and IPC, Fonda hoped to politicize moviegoers without preaching to them. The rhetorical strategy she found most effective was to "embed your passionate feelings into a delivery system that is totally accessible by people who do not agree with you. And that is what *The China Syndrome* does." While Richard's resistance comes too easy, as an almost Pavlovian response to authority by a stereotypical sixties protester, Kimberly and Jack become increasingly concerned and politicized as the narrative unfolds. Audience members are meant to take a similar journey, gradually coming to share Kimberly and Jack's passionate feelings about the dangers at the fictional plant.[15]

As Jack and Kimberly become more critical of the plant, they also become more emotional; while Richard's outbursts make him appear as an uninformed "hothead," their feelings appear legitimate. With their gradual transformation into dissenters and their concern based on fact and not just on fear and anger, they have earned the right to act as emotional citizens. As one reviewer noted, "Lemmon's convincing and sympathetic performance . . . is the film's linchpin, since his development . . . most closely parallels the audience's own—from a brooding sense of anxiety which grows into disillusionment and, finally, an overwhelming and inarticulate outrage at the recklessness of the energy corporation."[16]

The film melds fact with feeling in a crucial scene featuring two scientists who view the footage that Richard has secretly recorded. (In another blending of fact and fiction, one of these parts is played by Gregory Minor, a former nuclear engineer at General Electric who, along with two colleagues, resigned from the company in protest in 1976, claiming that "nuclear reactors . . . present a serious danger to the future of all life on this planet." The three engineers served as consultants to *The China Syndrome*.) After screening the footage, the two scientists tell Richard and Kimberly that the plant "came very close to the China Syndrome."[17]

"The what?" Kimberly asks.

One scientist then delivers the film's most famous lines: "If the core is exposed, for whatever reason, the fuel heats beyond core heat tolerance. . . . Nothing can stop it, and it melts right through the bottom of the plant, theoretically to China." He concludes with the line Gray borrowed from the Brookhaven Report: "The number of people killed would depend on which way the wind is blowing [and] render an area the size of Pennsylvania permanently uninhabitable."

The China Syndrome closes with a dramatic showdown at the nuclear power plant. Convinced that the company will sacrifice public safety for the pursuit of profit, Jack takes over the control room. A SWAT team then barges in and kills him. In the aftermath, one company official claims that Jack "had been drinking," while another describes him as "emotionally disturbed." But Kimberly repudiates these charges. Looking directly into the camera, she explains to viewers: "I met Jack Godell two days ago, and I'm convinced that what happened tonight was not the act of a drunk or a crazy man. Jack Godell was about to present evidence that he believed would show that this plant should be shut down." Barely holding back tears, she concludes: "I'm sorry I'm not very objective. Let's hope it doesn't end here." Back in the TV studio, the station managers rejoice at Kimberly's performance—shouting out "Spectacular!" and "She did a hell of a good job"—to indicate that her emotionality seems justified, not a sign of feminine weakness but an appropriate reaction to the tragic events. Just as *The China Syndrome* legitimates Jack's criticism of the plant and his increasing emotionalism, Kimberly's sexist managers now commend her affective display. Her feelings seem justified, an authentic expression of nuclear fear.

It is not surprising that the nuclear industry tried to discredit *The China Syndrome*, nor that conservative commentators lambasted its antinuclear message. Most notably, *Newsweek*'s George Will castigated the movie for its "anti-nuclear hysteria" and for projecting "propaganda" that was "parasitic upon public affairs." Will's review moved beyond the cinematic realm to mock the burgeoning antinuclear movement as a ridiculous throwback to the sixties: "For some people with an itch for moral action, life hasn't been the same since Vietnam, and anti-nuclear protest is a way of scratching the itch." Will wrote his piece just before the accident at Three Mile Island, and many readers subsequently wrote the magazine to excoriate the smug columnist. "Will Will take his crow roasted, broiled or fried?" one reader asked. "People who talk about safety standards for reactors in terms of 'trial and error' and 'risk analysis' statistics are idiots," another wrote. "Call it the George F. Will syndrome." According to these readers, the real-life accident refuted Will's willful ignorance and even proved, another letter writer claimed, that "Jane Fonda is usually right."[18]

What is surprising is that the film would receive some of its harshest criticism from the left. While Fonda and others associated with the film claimed to bring radical ideas to mainstream audiences, many reviewers in left-of-center periodicals found the politics of *The China Syndrome* severely wanting. Robert Hatch, writing in *The Nation*, began by praising the film for being "impeccably plotted and staged with . . . 'you are there' verisimilitude," but then questioned the film's "social effectiveness" and doubted whether it would play any meaningful role in shaping the debate over nuclear power: "The film, I regret to predict, will cause the utilities little trouble unless they go out of their way to make it. The picture's miscalculation, as an activist tract, is that its entertainment value is so high, and so similar in style to any number of other glossy melodramas with underpinnings of social concern . . . that its effect may well be to trivialize as passive entertainment a disturbing aberration of contemporary life."[19]

Murray Hausknecht, writing in *Dissent*, likewise wondered why the nuclear industry panicked about the movie: "There is nothing in the film that justifies assigning it so potent a role; it neither questions nuclear power systems nor the industry that runs them." For Hausknecht and like-minded critics, *The China Syndrome* failed to grasp the political economy of nuclear power and instead focused on evil individuals whose greed led them to sacrifice safety. The film, he concluded, "reflects an ideology that turns attention away from economic institutions to the behavior of 'bad' people and thereby presents the economic system as unproblematic, quite as much as the nuclear system."[20]

Hatch and Hausknecht challenged the conventions of Hollywood film for reducing all questions of power and politics to a dramatic story of individuals. Carrying their critiques even further, they suggested that the film's very status as a commodity, as a consumer item based on cinematic formula, limited its political vision. *The China Syndrome*, Hatch argued, would fail "to rouse the citizens from their lethargy," since its "varnished professionalism" did nothing more than "bath[e] them in the box-office tested pleasures of dazzingly realistic thrillers with socially pertinent overtones." Hausknecht directed his critique at Fonda as an embodiment of the Hollywood star system, someone whose "radicalism has been transformed by the movie into a commodity." This commodification of radicalism, he believed, produced "an illusion . . . that an important public issue has been examined from a critical perspective. Such illusions subvert genuine attempts to comprehend the problems of nuclear energy and undermine real impulses to radical action."[21]

These critiques may seem overly harsh and reductive, beholden to the same theories of mass culture that Gray, Fonda, and others sought to transcend in making *The China Syndrome* into a popular text enlivened with political purpose. Yet one does not have to accept the most sweeping condemnations of mass culture to acknowledge that Hatch, Hausknecht, and other leftist critics were onto something. It is not simply that the film was packaged as a gripping spectacle and that it relied on familiar conventions to entertain audiences. It is not simply that the film failed to offer a structural critique of capitalism and instead merely caricatured company executives as cartoonish embodiments of evil. Rather, *The China Syndrome* also narrowed its critique of nuclear energy to focus almost exclusively on the potential for a catastrophic reactor accident and, even more problematically, refused to portray citizens as having any capacity to shape the energy future.

Successive drafts of the script reveal this narrowing process. Early drafts refer to the unresolved problem of how to store nuclear waste, radioactive materials that "somebody is going to have to watch over . . . for the next twenty-four thousand years." Other drafts also situated the nuclear power debate within the broader context of the energy crisis and the effort to promote solar energy and soft-path solutions. Even though radioactive waste receives brief mention in the film, these questions about dominant energy systems and long-term concerns were largely removed the final screenplay. On narrative grounds this narrowed focus might be defended for streamlining the plot and enhancing the thrill factor of the film, making audiences feel an overwhelming sense of fear about the possibility of a major accident at a plant. Yet this limited focus also marginalized key components of the antinuclear critique and detached the nuclear power debate from broader energy politics.[22]

Even more disconcerting, as a number of reviewers noted, the film slighted, even mocked, the antinuclear movement, denying its potential to organize resistance. During scenes at a public hearing about the siting and construction of a new power plant, protesters are shown holding up photographs of their children. A female activist reads a statement: "These are pictures of children we love. . . . They, and others like them, will be the ones who will inherit the consequences of your actions. . . . Radiation is *not* healthy for children and other living things." This strategy of picturing children played an important role in the 1970s antinuclear movement. Indeed, these scenes which, according to Bridges's screenplay, featured interviews by Kimberly with "real people who are opposed to nuclear power" were intended to mimic actual hearings held in the mid-1970s about a proposed reactor in northern California's Diablo Canyon. The screenplay claims, in fact, that the lines recited by the female protester in the film were taken verbatim from a "real statement read at the hearings at Diablo."[23]

The protesters in the film, though, seemed problematic to many reviewers, who wondered why a film that mobilized concern about nuclear power would at the same time belittle antinuclear activists. "One curious aspect is the film's portrayal of the anti-nuke movement," a critic for *Cineaste* noted. "At a public hearing . . . anti-nuke activists . . . look irrelevant, not to mention downright silly. . . . It is unfortunate that the anti-nuclear movement comes off so poorly in a film which presumably shares its political perspective." In an interview, the director James Bridges not only acknowledged that the protesters looked foolish but also seemed to take perverse pleasure in the opportunity to deride them. "Yes, the meanies are painted black," he said, "but so are the anti-nukes, who come off ridiculous. They're standing there with their little photographs. . . . My favorite laugh in the movie is the anti-nukes standing there with their black handkerchiefs on their faces, because they're ineffectual."[24]

Even though *The China Syndrome* once again mixed fact and fiction, drawing on documentary evidence to depict actual protesters voicing authentic statements, the film still presented a selective, unflattering view of protest activity to make the demonstrators appear, in the words of one commentator, "over-dramatic, silly, and . . . uninformed." Another critic explained: "The film ignores or belittles the impact of collective action. Instead of an organized grassroots anti-nuclear movement, the movie presents a bunch of 'no-nukes kooks' straight out of the media stereotype films." As critics indicated, the cinematic protesters use only their feelings to convey their concern and thus seem, much like Richard, the radical cameraman, to be "hysterical," acting beyond the emotional framework deemed acceptable by the film.[25]

The film ignored the link between cognition and emotion in the antinuclear movement. Members of such groups as the California-based Mothers for Peace—supposedly the model for *The China Syndrome*'s protesters—emphasized their role as nurturing mothers seeking to protect their children and futurity from harm. Even as Mothers for Peace and other antinuclear groups engaged in emotional protest, they fused fear with fact, combining scientific knowledge with parental concern to articulate their dissent from nuclear power. By showing the activists briefly, and depicting only the affective dimension of their protest strategy, *The China Syndrome* marginalized the ongoing struggle against nuclear power and made the movement appear meaningless in public life.[26]

A more charitable reading of these scenes, one that Fonda later tried to promote, would insist that the filmmakers wanted to portray the undemocratic nature of nuclear hearings, to reveal that the licensing process was deliberately designed to preclude full public participation. Rather than fostering meaningful citizen involvement, the hearings produced instead a democratic simulacrum that effectively shielded the industry and government from critics. Fonda claimed that activists understood exactly what these scenes aimed to accomplish and that the filmic hearings resonated with their own experiences of being ignored by managerial elites.[27]

While the scenes do indeed suggest the challenges activists faced, Fonda's reading ultimately seems unconvincing. *The China Syndrome* presents a decontextualized vision of the antinuclear movement: a small group of activists show up at some hearings, hold up pictures of their children, and, after being shunned and silenced by the authorities, stage a puny protest outside and never appear again in the film. Their brief, ineffectual presence provides viewers with no context to understand the emergence of this movement, to appreciate the potential impact of citizen protest, or to recognize the actual coalitions being formed at the time of the film's release between dissident scientists and activist groups like Mothers for Peace. While Gray's script referred to an organization called Scientists for the Environment (clearly modeled after the Union of Concerned Scientists), in Bridges's drafts and in the film antinuclear scientists are depicted as lone dissenters; they remain unorganized and, like the female protesters, completely ineffectual.[28]

In their critiques of *The China Syndrome*, many leftist reviewers seemed to pine for what the film theorist Jane Gaines terms "political mimesis." A common feature of documentary films, political mimesis often involves scenes of protest that move audiences to take action, that encourage them "to do something instead of nothing in relation to the political situation illustrated on the screen." During the 1970s, antireactor activists produced a number of documentary films that sought to publicize the dangers of nuclear power and to model collective political action. These documentaries presented organized resistance and group protest as crucial expressions of citizenship. As an activist explained in one film, "It's our duty as citizens to oppose and protest what we believe is dangerous and unresolved technology." Other documentaries featured extensive footage of marches, rallies, and civil disobedience meant to inspire citizen action in the face of environmental danger. Even though *The China Syndrome* included protest scenes, leftist critics believed that this imagery did not constitute an inspiring form of political mimesis and thus would not move audiences to act. Neither conservative nor radical critics gave audiences much credit: while conservatives worried that spectators would be duped by Hollywood spectacle, leftists complained that the film entertained rather than politicized.[29]

But while leftist critics railed against *The China Syndrome* for being insufficiently radical, antinuclear activists embraced the film as a propaganda tool to educate the public and stir action. Indeed, almost immediately upon its release, activists started to gather outside movie theaters, armed with informational leaflets to distribute after each showing (fig. 8.3). Hatch, Hausknecht, and other leftist reviewers seemed to imagine audiences as being fully immersed in the action on the screen, unable to make connections to the world outside the cinema. Their harsh judgments about the inability of Hollywood to convey subversive ideas rang hollow to activists who sought to harness the film to advance their cause. Some would find "a new openness, even hunger for information," and that many filmgoers were "anxious to take leaflets" and "to discuss the issues raised in the film." All across the United States, as well as in Canada and the United Kingdom, antinuclear demonstrators distributed leaflets, urged spectators to write

FIGURE 8.3. Antinuclear protester displays a sign underneath *The China Syndrome* cinema marquee, Philadelphia. Photograph by Ed Eckstein, 1979. Copyright Ed Eckstein / Corbis.

government officials, and tried to recruit members into their organizations. In San Luis Obispo, California, the Abalone Alliance—another group opposed to the Diablo Canyon reactor project—staged a march that began at the local theater playing *The China Syndrome* and ended at the regional office of PG&E, the utility company that planned to build the reactor. During the march, Abalone members carried a coffin to signify the danger to vulnerable ecological bodies. While the Diablo hearings had provided a template for the film, now real-life protesters used the movie to heighten public visibility of their struggle. The mass culture text, the commodity maligned by leftist critics, became not merely a spectacle for consumption but a powerful means to politicize nuclear fear and promote the right of citizens to be free from the risk of reactor meltdown.[30]

The China Syndrome, despite its unflattering portrayal of activists, became an activist tract and even, some claimed, a cinematic prophecy that seemed to prepare audiences for the reality of Three Mile Island. Before March 16, only a handful of people had known what the title meant. Twelve days later, as they tuned into TV coverage of the accident, many began to wonder whether the China syndrome would come true.

NINE

Nuclear Meltdown II: Three Mile Island

On March 28, 1979, Walter Cronkite, the most revered news anchor in television history, offered CBS viewers a grave gloss on the day's events at Three Mile Island: "It was the first step in a nuclear nightmare. As far as we know at this hour, no worse than that. But a government official said that a breakdown in an atomic power plant in Pennsylvania today is probably the worst nuclear reactor accident to date." Cronkite stressed the gravity of the situation but also assuaged audiences by confirming that the crisis had not devolved into full-scale catastrophe: "Plant officials said today's accident was not a China syndrome situation—referring to the current movie about a near-catastrophic nuclear meltdown."[1]

From the first day of the accident, *The China Syndrome* defined the parameters of concern: Three Mile Island would be compared with the cinematic text, providing audiences with a way to understand the severity of the crisis and to contemplate whether it might become a worst-case scenario. That night, all the network news broadcasts summarized the technical aspects of the accident: the breakdown of a cooling pump, the buildup of heat and pressure, the release of radioactive steam. They also ominously noted that radiation was still "leaking out through the walls" and "was detected as far as one mile away." News broadcasts left audiences wondering if the conditions might worsen in the hours and days ahead.[2]

According to the film theorist Mary Ann Doane, television's emphasis on "the potential trauma and explosiveness of the present" plays a crucial role in shaping popular understanding of events. In this light, we can see the media framing of Three Mile Island as coverage that organized time in an emotionally powerful manner: over the course of several nights, TV news broadcasts presented the accident as a crisis that potentially could become a catastrophe. The crisis, Doane explains, "names an event of some duration which is startling and momentous precisely because it demands resolution within a limited period of time." Each night, as anchors and correspondents provided updates on the accident, the question—sometimes posed explicitly, other times hovering implicitly—would be whether officials could resolve the crisis and prevent a catastrophe—a core meltdown and China syndrome—from happening. News coverage provided audiences with a feeling of ongoing, extended involvement with the crisis, with a sense of living through a tense period in which at any moment "the punctual discontinuity of catastrophe" and the terror of "technological collapse" might occur.[3]

After the crisis ended, photographs of Three Mile Island's cooling towers became icons of the accident, visual reminders of the tense moments that gripped the nation. Television and still photography placed viewers in a different temporal relationship with the accident: while TV news generated a sense of expectant urgency as the crisis unfolded, news photographs stopped time to offer a retrospective depiction of the site. Although they operated in different temporal modes, both TV news and still photography mobilized public fear and helped validate the emotional politics of the antireactor movement.

Nevertheless, even as they popularized antinuclear concern, these images also detached the accident from broader manifestations of energy crisis and focused public attention on the nuclear power plant as the sole locus of environmental danger. Media temporality ignored the slow violence of the nuclear fuel cycle, including the experience of nuclear colonialism by indigenous communities. This coverage also failed to consider the long-term visions of sustainability developed by alternative energy advocates, and slighted their effort to incorporate hope into the emotional framework of American environmentalism. While the short-term, immediate danger of the China syndrome and Three Mile Island became increasingly visible in American public culture, these more extensive time frames of fear and hope remained marginal to spectacle-driven media framings of environmental crises.

On March 30, the third night of the accident, Cronkite began his broadcast with this shocking summary: "The world has never known a day quite like today. It faced the considerable uncertainties and dangers of the worst nuclear power plant accident of the atomic age. And the horror tonight is that it could get much worse." Concentrating on this moment as particularly frightening and a harbinger of potential calamity, all the news programs that night explained that even the Nuclear Regulatory Commission—which, up until then, had downplayed the danger of Three Mile Island—now acknowledged that a meltdown might occur. A hydrogen bubble had formed inside the reactor, and, as the size of the bubble increased, the situation became more critical. "It is not an atomic explosion that is feared," Cronkite continued. "The experts say that is impossible. But the specter was raised of perhaps the next most serious kind of nuclear catastrophe—a massive release of radioactivity. The Nuclear Regulatory Commission cited that possibility with an announcement that, while it is not likely, the potential is there for the ultimate risk of a meltdown."[4]

That night, all three networks provided extensive coverage of the crisis, with ABC and CBS both devoting more than half of their news time to Three Mile Island. In addition to the hydrogen bubble story, they also reported on the Pennsylvania governor's announcement that all pregnant women and young children should evacuate the area. As the hydrogen bubble provoked visions of catastrophe and as radiation continued to leak from the plant, TV news emphasized the risk to human health, especially the vulnerability of fetal and children's bodies. This coverage reinforced the antireactor movement's sense of urgency and made nuclear power appear as an imminent threat to the citizenry. TV news amplified the sense of danger and invited audiences to glimpse an accelerating crisis that might burst into catastrophe.[5]

Media reports repeatedly referenced *The China Syndrome* to understand Three Mile Island. ABC's *20/20* news program, for example, interspersed scenes from *The China Syndrome* with news footage of Three Mile Island. "Some of the parallels between the movie and the events in Pennsylvania are almost too close for comfort," the program concluded. Likewise, *Newsweek* magazine included a film still—a picture of Jane Fonda standing in front of a diagram of a nuclear power plant—and then detailed some of the "striking" similarities "between the film and the drama at Three Mile Island." For all the similarities, though, *Newsweek* noted, "the real events proved more frightening. . . . In the movie, there is never any question of radiation leaking—as it did in real life." The film, another commentator explained, "didn't have women and children being evacuated.'"[6]

While all three networks showed Three Mile Island's cooling towers as prominent features of the landscape, only ABC repeatedly framed the structures as signs of technological danger. "Visuals of the nuclear plant possessed a Gothic quality," two communication scholars observed, "especially on days when ABC correspondents did stand-up reports with the plant's massive cooling towers, enveloped in mist, looming in the background." Forbidding towers contrasted with bucolic scenes as ABC frequently cut from the reactor to "panoramic views of farms, cattle grazing in fields, and school buses departing the area." These reports also emphasized the towers' threat to futurity, as they depicted children as being particularly vulnerable to radioactive danger. After milk samples from nearby dairy farms tested positive for iodine 131, ABC powerfully conveyed the risk to young bodies and the local ecology. The report showed cows grazing while the cooling towers hovered in the background. Meanwhile, audiences heard "the cry of a baby and the crackling of a Geiger counter."[7]

After the Three Mile Island crisis ended, photojournalists and other image makers repeatedly depicted the towers as symbols of nuclear danger, turning the structures into an environmental icon. *Time, Newsweek*, and other popular magazines ran cover stories with such titles as "Nuclear Accident" and "Nuclear Nightmare" that featured ominous-looking photographs of Three Mile Island's cooling towers (fig. 9.1). The concrete verticality of the towers—rising more than 350 feet above the Susquehanna River—offered the most striking visual emblems of the plant. Photographers used zoom lenses to detach the towers from much of their surrounding environment, to make them appear as enormous, threatening structures that dominated the landscape. The photographs on the *Time, Newsweek*, and *Life* covers were all taken at either dusk or dawn to accentuate the eerie loom of the towers. Magazine covers relied on color photography to reveal another menacing feature: red lights blinking on the towers, heightening the sense of danger and potential catastrophe.[8]

While the cooling towers became an environmental icon, the actual accident at Three Mile Island occurred in the far more mundane-looking containment building, a cylindrical structure that surrounded the reactor and that stood, more modestly, at about half the height of the towers. In the *Time* cover image, the containment structure can be

FIGURE 9.1. *Time* magazine cover, April 9, 1979. Copyright 1979 Time Inc. Used under license. *Time* and Time Inc. are not affiliated with, and do not endorse products or services of, licensee.

seen in the background, looking minuscule behind the massive towers, while the *Newsweek* photograph completely ignores the building.

It is worth noting that the cooling towers at Three Mile Island and elsewhere actually represented a concession by the nuclear industry to the environmental and antinuclear movements. In the late 1960s, environmentalists began to protest thermal pollution—the degradation of water quality caused by waste heat from nuclear power plants. In response, "the nuclear industry gradually and reluctantly took action . . . by building cooling towers . . . for plants that lay on inland waterways." Since Three Mile Island's two reactors were built after the thermal pollution controversy, each came equipped with two cooling towers.[9]

While some news photographs showed all four towers situated amid a wider view of the island, the most famous images focused closely on one or two towers, especially at dusk when the serenity of the sky seemed to be overtaken by the sinister structures. No longer banal features of the local landscape, the towers became enduring reminders of the near catastrophe at Three Mile Island. Like images of the *Hindenburg* explosion, the *Challenger* space shuttle explosion, and other icons of technological disaster, the photographs conveyed "a sense that danger lurks beneath the surface of a supposedly safe, controlled, modern environment." The images provided a concrete representation of Three Mile Island's hidden danger—even if that danger primarily lurked in a different building.[10]

Many news photographs, editorial cartoons, and posters portrayed the towers not simply as abstracted, isolated forms but as being in dialogue with people and other objects to amplify the emotionality of the scene. Taken by Martha Cooper, a widely circulated Associated Press photograph shows a white woman, presumably a mother, holding the hand of a white child (fig. 9.2). Standing in the foreground, perfectly positioned between the two cooling towers in the distance, the mother offers protection to the toddler. In the middle distance plowed fields, widely spaced homes, and mature trees suggest a setting that should nurture and sustain local families. Yet the towers negate the pastoral vision and signify the possibility of a technological collapse that could endanger residents, especially children. The proximity of mother and child to the towers raises troubling questions about the dangers to futurity, and activates concern for the protection of vulnerable bodies.

The theme of childhood vulnerability appears in another AP photograph, taken by Barry Thumma on March 30, the day the governor of Pennsylvania called for the evacuation of pregnant women and children (fig. 9.3). In the image, a single cooling tower looms over an empty swing set. The juxtaposition of swing set and cooling tower frames nuclear energy as potentially harmful to children, and thus reinforces the emotional politics of the antireactor movement. Much like SANE ads, gas mask pictures, and other images of children at risk, this

143

FIGURE 9.2. Mother and daughter near cooling towers, Three Mile Island. Photograph by Martha Cooper, 1979. AP Photo / *Harrisburg Patriot-News* / Martha Cooper. Used by permission of AP Photo.

photograph presents the abandoned playground as a site of childhood innocence now threatened by environmental danger.

The continued circulation of the cooling towers in other visual media demonstrated their iconic status. In the aftermath of the accident, a number of leading editorial cartoonists appropriated cooling tower imagery to critique nuclear power. For example, Paul Conrad's "The Unthinkable" portrayed Rodin's *The Thinker* as petrified instead of pensive, looking aghast as he envisions a cooling tower exploding and taking the form of a mushroom cloud (fig. 9.4). The two icons of the nuclear age—the cooling tower and the mushroom cloud—merged to produce an apocalyptic warning about the dangers of nuclear energy. Like other media depictions of Three Mile Island, the cartoon emphasized the sudden temporality of catastrophe rather than the slow violence of environmental time. While SANE and other antinuclear activists had moved beyond the mushroom cloud to visualize the accumulating danger of fallout, popular portrayals of Three Mile Island relied on the punctuated experience of disaster to represent the hazards of this energy system.[11]

Meanwhile, *Mad* magazine posed the freckle-faced, usually nonchalant Alfred E. Neuman in front of two cooling towers (fig. 9.5). Radioactive material, its vivid redness echoed in the tower's warning lights and jagged crack, bursts from the top, again signaling the risk of sudden meltdown. Devoid of his famous grin and with beads of sweat on his face, Neuman revises his famous line, "What, Me Worry?" to announce, for the first and only time in the magazine's history, "Yes . . . Me Worry!"[12]

Before Three Mile Island, cooling towers were not automatically associated with nuclear danger. Indeed, *The China Syndrome*, despite its effort to achieve visual verisimilitude, did not depict cooling towers in the film or in its promotional materials. What makes this omission all the more surprising is that the film's production designer George Jenkins, who received an Oscar nomination for his work on the film, visited the Trojan Nuclear Power Plant in Oregon. Traveling with James Bridges and Michael Douglas, Jenkins studied the design of the reactor

FIGURE 9.3. Three Mile Island cooling tower looms behind an abandoned playground. Photograph by Barry Thumma, 1979. AP Photo / Barry Thumma. Used by permission of AP Photo.

THE UNTHINKABLE

FIGURE 9.4. "The Unthinkable," Editorial cartoon by Paul Conrad originally published in the *Los Angeles Times*, 1979. Used with the permission of the Conrad Estate. This item is reproduced by permission of the Huntington Library, San Marino, California.

and used its control room as a model for the movie. Nevertheless, even though the Trojan plant boasted a massive cooling tower—standing at 499 feet, almost 150 feet taller than the Three Mile Island towers—*The China Syndrome*'s plant exterior did not feature a cooling tower but instead consisted of a less imposing, dome-shaped containment building. No advertisements, no publicity stills, no shots in the movie gave any indication that the filmmakers or marketers found apocalyptic meaning in cooling towers.[13]

Media coverage of Three Mile Island changed that. In 1999 and

2004, to coincide with the twentieth and twenty-fifth anniversaries of the film, the DVD cover prominently displayed an image of a cooling tower below pictures of Jane Fonda, Jack Lemmon, and Michael Douglas (fig. 9.6). The cooling tower was now being appropriated to market and package a fictional film. The blurring of Hollywood and reality continued.

FIGURE 9.5. Back cover illustration by Norman Mingo, *Mad*, October 1979. From MAD #210 ™ and Copyright E.C. Publications Inc.

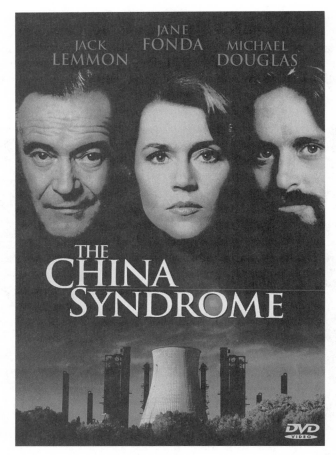

FIGURE 9.6. DVD cover, *The China Syndrome*, 1999. Copyright 1999 layout and design Columbia TriStar Home Video. All rights reserved. Courtesy of Columbia Pictures.

The Three Mile Island crisis altered dominant views of the antinuclear movement. Throughout the 1970s, grassroots groups across the United States—the Clamshell Alliance in New England, the Abalone Alliance in California, and scores of other local coalitions—organized, protested, and engaged in massive acts of civil disobedience to oppose nuclear plant construction in their communities. Although they rarely succeeded in halting projects altogether, their efforts helped delay the construction process and, in turn, increased the cost of this already highly expensive, capital-intensive energy technology. Even though *The China Syndrome* marginalized the role of protesters, these grassroots movements—along with national environmental groups that joined the antinuclear cause—contributed to growing public doubt about nuclear power. While the mainstream media had previously downplayed and denigrated the actions of these groups, the dominant frames changed considerably following Three Mile Island.[14]

On May 6, 1979, antinuclear groups organized what at the time was the largest protest against nuclear power in US history. Between 65,000 and 125,000 demonstrators marched on Washington, chanting, "Two-four-six-eight, we don't want to radiate." As with previous antinuclear rallies, the media drew comparisons between the marchers and 1960s protesters. On ABC, correspondent Bill Zimmerman began his report by observing, "It is reminiscent of the more peaceful demonstrations of the 1960s." *Time* included a photograph of a long-haired male protester outfitted in a gas mask and holding a sign that proclaimed, "Hell No, We Won't Glow." The caption read, "A touch of decades past," while a brief quote from a demonstrator explained, "I imagine this is what the 60s were like." The sixties references continued as almost every media report emphasized the presence of movement luminaries who addressed the crowd that afternoon, including Dr. Benjamin Spock, Jane Fonda, and her husband Tom Hayden, the former leader of the Students for a Democratic Society. "On the Capitol steps," *Newsweek* noted, "Ralph Nader, Tom Hayden and Jane Fonda summoned up memories from the '60s and monstrosities from the China Syndrome."[15]

While the media had previously applied the sixties frame to dismiss the antinuclear movement as a marginal gathering of misguided hippies, disaffected radicals, and romantic outcasts with no scientific credentials, now the so-called "60s spirit" acquired a more positive dimension. After ABC's Zimmerman described the crowd as representing "the old peace coalition with some new additions," the camera showed a young white child being pulled in a red wagon. Likewise, NBC interviewed a white female participant who marched alongside a white man and commented, "I am very concerned about the welfare of our children." A later clip showed the man pulling their two children in a red wagon. Even as this coverage sometimes cast the antinuclear movement as nothing more than "a nostalgic search for a new issue," composed of "graduates of an earlier era returning for a 10-year reunion," the media also accorded new respect to these sixties protesters, emphasizing that they were now parents and were thus justifiably concerned about their children's future. By showing images of parents and children together, TV and other media sources lent greater legitimacy to the movement and depicted it as a cause committed to futurity. While *China Syndrome* director James Bridges, together with many critics, found the filmic protesters' use of children's photographs to be ineffective, the news media portrayed these marchers as earnest, caring parents seeking to protect the fragile bodies of children from radioactive harm.[16]

In sweeping terms, media reports also described the demographics of antinuclear protesters: "mostly middle-class" and "virtually all . . . white." The whiteness of the protesters, their putative privilege, and the focus on innocent white children as potential victims of nuclear catastrophe followed other mainstream framings of environmentalism, from the SANE campaign against radioactive fallout through gas mask imagery. The overwhelming presence of white, middle-class protesters once again reinforced the idea of universal vulnerability: the notion that all Americans, no matter their race or class background, could be threatened by nuclear danger. Yet in this case the media also implied a certain narrowness in the antinuclear movement's constituency. "The poor and minorities," one report noted, were "noticeably lacking" from the protests.[17]

Although the media commented on the whiteness of the movement, this same coverage failed to depict the risks nuclear power posed to other groups. On July 16, exactly four months after *The China Syndrome*'s release and just over three months after Three Mile Island had dominated news headlines, the worst radioactive spill in US history occurred near Church Rock, New Mexico. On lands held by the Navajo Nation, the United Nuclear Corporation mined and milled uranium to supply fuel to nuclear power plants and, in the process, often disregarded the health and environmental effects on indigenous communities. That morning a tailings dam broke, releasing ninety million gallons of liquid radioactive waste and more than one thousand tons of tailings solids into the nearby Rio Puerco. "The Rio Puerco accident . . . far outweigh[ed] the well-reported Three Mile Island accident," the sociologist Valerie Kuletz observes. "Why wasn't this massive radioactive pollution reported in the national press?" Following the spill, she continues, "the Navajo people in the surrounding area were unable safely to use their single source of water, nor could they sell or eat the livestock that drank from this water. Acute toxicity cased by increased acidity of the water resulted in burns leading to infections that required amputations."[18]

Despite the severity of the spill, this accident would be almost completely ignored by television news and other media outlets. Even though the three major networks provided extensive coverage of Three Mile Island, on some nights devoting more than half of their broadcast time to the crisis, Rio Puerco would barely register a blip on TV news. Indeed, the spill was not even mentioned until several weeks later, when ABC, the only national network to cover the catastrophe at all, aired a single report on one nightly news broadcast. The segment offered one Navajo leader the opportunity to voice his criticism of the government and media's neglect of the disaster that was "equal to Three Mile Island in our minds. . . . If it happened [elsewhere], you better believe there would be a lot of action done" in response, he asserted.[19]

Rio Puerco's paucity of coverage could be attributed to the Navajo Nation's relative geographic isolation far from dominant urban and media centers, and to the ongoing neglect of present-tense Indians—as opposed to the Crying Indian and other past-tense natives—in American public culture. But it is also the case that the Rio Puerco disaster did not involve an iconic cooling tower. *The China Syndrome*, media coverage of Three Mile Island, and popular framings of the antinuclear movement all focused on the reactor as the primary site of concern. The presumed victims could easily be imagined and depicted as white children, those familiar emblems of universal vulnerability, being pulled in red wagons at protests or living in the vicinity of nuclear power plants. The exclusive emphasis on the reactor as a site of danger obscured all the other spaces in which nuclear power affected the lives and landscapes of different groups of Americans. Indeed, the ABC segment on Rio Puerco failed to mention that uranium from the area ended up in nuclear reactors, perhaps even at Three Mile Island. The connections between these disparate places—the material links between uranium mining and energy production—remained hidden from view. Beyond the Rio Puerco disaster, the other long-term legacies of nuclear colonialism— the accretive damage to human bodies and ecosystems caused by uranium mining—failed to gain popular visibility.[20]

Likewise, even as the media's sense of crisis fostered concern for children's health and the protection of futurity, this focus on immediate danger ignored the longer time scale of toxicity represented by nuclear waste. Just as iconic cooling towers and meltdown fears worked to obscure the geographies of uranium extraction, media images neglected the spatial and temporal dimensions of radioactive waste. Power production at Three Mile Island and elsewhere connected reactors to distant landscapes and communities, to diverse histories, and to unimaginable futures. But popular portrayals of nuclear power obscured these links between different groups of people and environments, these relationships forged in space and time, and these extensive time frames of risk.

According to conservative commentators, *The China Syndrome* and Three Mile Island produced antinuclear hysteria and turned the American public against a clean, safe energy source. Hollywood film and news media spectacle worked together to manipulate audiences into accepting the claims of the antireactor movement. In the process, these critics argue, "stark and simple images have replaced complex realities." This conservative narrative presents the decline of the nuclear industry as a story of emotion overpowering reason, of spectacle displacing thought, of image misrepresenting reality.[21]

Yet such a reading simplifies a more complex historical record. As a number of scholars have noted, the nuclear industry's troubles began even before the Three Mile Island accident, as "high interest rates, massive cost overruns, lengthy construction delays, public protests, and, in some cases, outrageously shoddy construction combined to force the cancellation of many plants." The public feelings unleashed by *The China Syndrome* and Three Mile Island reinforced these economic trends to exacerbate the industry's problems and to make new plant orders appear almost "inconceivable." Culture and structure, public feelings and economic reality combined to discredit the industry. While Project Independence and other energy proposals envisioned the construction of hundreds of new reactors, nuclear power proved to be an "economic disaster."[22]

By assuming that media images helped environmentalists get exactly what they wanted, the conservative narrative also obscures the full range of antinuclear and alternative energy ideas that were, for the most part, ignored at the time. In addition to masking the long-term risks of uranium extraction and radioactive waste, media framings of energy issues slighted the movement's long-term social vision. While environmentalists sought to move beyond the politics of negation, to promote a collective vision of hope for the future, their efforts rarely gained positive attention from the spectacle-driven mass media.

About six months after *The China Syndrome*'s release, Jane Fonda addressed a crowd of approximately two hundred thousand gathered in lower Manhattan for the largest antireactor event ever held in the United States. Organized by the Musicians United for Safe Energy (MUSE), the rally and concert closed a week of "No Nukes" concerts. In her speech Fonda, like the MUSE leaders, sought to connect antinuclear protest to the struggle for alternative energies. Standing at a podium adorned with a pro-solar banner, Fonda hailed renewable energy as a part of a broader vision to improve American life. "We need an energy source" such as "solar, which is job-intensive, which is democratic, which is decentralized, which is safe," she announced. Fonda's emphasis on job creation and on the democratic and decentralizing promise of solar energy encapsulated some of the key claims of the alternative energy movement. While media reports showed antinuclear protesters carrying banners that announced "No Nukes" and "Turn On the Sun" or holding Frisbees with an image of the sun and the words "We Don't Need Nuclear Power," this coverage almost never explained the larger, collective vision of hope that inspired many environmentalists. While the short-term spectacle of nuclear meltdown furthered the reactive politics of risk, the long-term effort to refashion the nation's energy future did not conform to dominant patterns of media temporality.[23]

Ultimately, *The China Syndrome* and Three Mile Island demonstrated how the media fragmented popular perceptions of the energy crisis and rendered incoherent the claims of oppositional movements. Even as the media warned of technological catastrophe and of the dangers radiation posed to children and the unborn, these same channels of communication rarely took seriously the idea that alternative energy activists constituted a movement with a promising vision of the future. Even as Fonda and others sought to connect the fear of meltdowns with the hope of renewable energy, media temporality—the focus on immediate crisis and the failure to engage with longer time scales—worked against alternative energy advocates. Solar energy seemed alluring, but the soft path never gained any traction.

TEN

Here Comes the Sun?

"On May 3, 1978, the Solar Age will begin!" Denis Hayes announced. "Mark it on your calendar. Talk about it. Think about it. Make plans. Get ready for SUN DAY . . . the international demonstration of humankind's passage from the Petroleum Age to a new era of unlimited, decentralized, economical, pollution-free, ecologically sound energy use . . . *the era of solar energy!*" Clearly, Hayes had high hopes for Sun Day. The national coordinator for the first Earth Day, he believed that Sun Day could be just as successful in gaining media attention and mobilizing public support for a sustainable energy future.[1]

Sun Day organizers depicted the dawning of the "Solar Age" as a sign of hope, an encouraging glimpse into the future. According to Peter Harnik, Earth Day veteran and outspoken critic of the Ad Council's incessant focus on individual responsibility, solar energy could unify diverse groups of Americans around a collective vision of change. For this reason, Sun Day provided a chance to reframe the emotional politics of environmentalism. "It should be a powerful answer to those who've been saying that environmentalists are just prophets of doom and don't have a positive program," Harnik explained. Rather than focusing on apocalyptic visions and doomsday scenarios, the alternative energy movement sought to mobilize feelings of hope. As Hayes put it, while "Earth Day emphasized a problem, Sun Day emphasizes a solution."[2]

Sun Day's visual logo drew on the familiar notion of the sun as an emblem of hope to express a humanistic vision

FIGURE 10.1. Sun Day logo by David Root, 1978.

of solar energy (fig. 10.1). The image depicted people joining hands, uniting together to constitute the sun's rays, forming a larger sense of community around the luminous celestial body. Solar energy represented not only an image of enlightened technology and future promise, but also a way to restore the nation's civic vision, to inspire and unify during a time marked by pessimism and declining faith in social institutions. "Sun Day provides a new glimmer of hope and optimism after a period of diminishing expectations," Harnik noted. "Whereas Earth Day awakened us to a whole set of environmental problems that had been ignored for decades, Sun Day will celebrate positive, attainable solutions." The Sun Day logo presented solar energy as a unifying cause that could galvanize the citizenry by focusing public attention on "the people's energy source."[3]

Sun Day sought to emphasize the collective meanings of alternative energy. Yet this broader vision would be dismissed and even mocked in media depictions of the event. Indeed, Sun Day's collective vision of hope did not fit with the dominant media frames applied to the alternative energy movement during the 1970s. While the Sun Day logo pictured the prospect of a renewable future and dreamed of the democratization of energy policy, media portrayals of alternative energy instead almost always focused on individual tinkerers and homeowners. The quest for a solar future thus appeared as a project carried out by individual Americans rather than a broader movement to reorient the nation's energy policy. Much like the "Don't Be Fuelish" campaign, these depictions of alternative energy emphasized the virtue of personal action. The idea that solar might be part of a larger struggle to decentralize and democratize the United States, that it might lead to a better quality of life and encourage greater citizen involvement in public affairs, would be completely absent from this coverage. Media reports repeatedly emphasized stories of lone inventors and solar pioneers, but denied any larger social vision and ignored the collective dreams that inspired many renewable energy advocates.

TV and other media sources also disregarded the temporal politics of environmentalism. According to alternative energy activists, the present moment represented a time of passage, a period when the United States should begin taking the soft path of conservation and renewable energy rather than following the hard path of petroleum, nuclear, and other nonrenewables. While the sense of immediate crisis fostered by *The China Syndrome* and Three Mile Island had helped popularize one aspect of the antinuclear critique, the media failed to consider the longer temporal perspective offered by Sun Day organizers and soft path advocates. Even on Sun Day—in fact, especially on Sun Day—this coverage reduced the movement to a smattering of individual tinkerers and romantic dreamers who hitched their hopes to the sun but failed to engage with larger struggles over power and politics.

It would be too simple, though, to assign all the blame to the media for this narrow vision of Sun Day and the soft path. Indeed, event organizers—along with the alternative energy movement more generally—too often presented solar energy in depoliticized terms and promoted a form of techno-determinism that imagined this energy source automatically ushering in a more democratic, decentralized society. Sun Day organizers wanted to frame solar energy as a consensual cause, "an apple pie issue" that everyone could support. "The whole thing is billed as the most upbeat and noncontroversial pro-environment campaign in many years," one commentator noted. "Sun Day isn't *against* anything—it's *for* sun power." Through this apolitical, nonthreatening approach, Sun Day organizers sought to broaden public support and curry media favor. Yet this effort to downplay power and politics ultimately diluted both Sun Day and the soft-path struggle. As one environmentalist noted at the time, Sun Day became "bizarrely apolitical" and "frivolous." Rather than challenging corporate power, rather than questioning the growth imperative, rather than connecting environmentalism to other social movements, the day became nothing more than a series of "sundae discounts, hot-air balloon launchings, and sunflower paintings."[4]

Sun Day organizers did not directly challenge dominant energy systems, but instead packaged solar as a sustainable source that deserved to become a larger part of the nation's total energy mix. As Denis Hayes explained in a Sun Day interview with CBS News, "We are a broad, diverse group who share in common one relatively narrow objective: to try to get an even break for solar energy." Even as Hayes and other energy activists recognized the ecological limits to growth, even as they critiqued the long-term dangers of the hard path, and even as they forged alliances with labor and other social movements, they decided to tone down their political analysis to fashion Sun Day as a depoliticized "celebration of the sun." They believed that this event would rival, perhaps even surpass, Earth Day in its impact upon environmental politics. Yet Sun Day's effort to mobilize collective hope failed to make a lasting imprint on American life. The event was bedeviled by its own contradictions and by media frames that emphasized the individualist meanings of alternative energy.[5]

Throughout the 1970s, TV news and popular magazines presented positive accounts of solar energy. These reports usually depicted new solar-powered homes being built, or old homes being retrofitted to take advantage of the sun's energy. From technical discussions of photovoltaic cells to financial discussions of the savings homeowners would find in lower energy bills, this coverage focused on solar as a form of scientific experimentation, and on the economic costs and benefits experienced by individual Americans.

In a story that demonstrated the emphasis on individual tinkerers, ABC News introduced viewers to Harry Thomason, an inventor and patent attorney who had recently built a solar-heated home in a suburb near Washington, DC. An ABC crew joined government officials who took a tour of Thomason's house and marveled at his ingenuity. "Those who come to [his] house usually leave believers," the correspondent explained. Thomason was not portrayed as some back-to-the land hippie enduring spartan conditions or suffering amid self-imposed deprivation. Not only did his family live in what the reporter described as "a conventional-size house"; the footage showed his wife and children frolicking in their heated indoor pool. Even though ABC cameras trailed politicians and federal officials—people, that is, responsible for the making of energy and environmental policy—the tour of Thomason's home was framed entirely as the story of one solar pioneer, not as an exemplar of a fundamental change in the nation's energy system. One member of the tour group, Representative Mike McCormack, reinforced this view: "Of course what he has done is demonstrate that it can be done," McCormack said. "And it can be done inexpensively. And this provides a great deal of incentive for other people to go out and try their hand at it, try their own innovations at it."[6]

This do-it-yourself, individualist frame would continue to be applied to solar stories throughout the decade. Jules Bergman, the science editor for ABC News, personalized the solar issue in an unusual manner: he brought a camera crew to his own suburban home in Palisades, New York. "After reporting on alternative energy stories for years," Bergman remarked, "I decided to install solar units myself for water heating." Pleased to join the ranks of solar pioneers, Bergman reported that the new system did not "take up much space." As an added bonus, he expected to recoup his "investment in four years." Bergman thus presented himself not as a starry-eyed idealist or a holier-than-thou environmentalist, but rather as a practical homeowner and informed consumer, someone who had crunched the numbers and believed that, for him, solar water heating made economic sense.[7]

The visual depictions of Thomason and Bergman's homes emphasized their suburban settings: these were detached domiciles that appealed to their owners' sense of self-sufficiency. The autonomous homes represented a self-contained vision of solar change, an expression of individual freedom and enlightened consumerism. This focus on single-family dwellings in the suburbs presented solar as a potential panacea to the energy crisis. Yet these portrayals also isolated solar from broader, collective visions of sustainability and ignored the environmental costs of suburban sprawl. As two architecture critics noted at the time, the individualist, techno-frame threatened to turn solar into "a mechanism to perpetuate . . . inefficient patterns" of fossil-fuel dependent growth, "rather than a means to a more environmentally sound culture."[8]

In contrast, many environmentalists and alternative energy advocates placed solar in a broader frame to imagine it as part of a fundamental restructuring of the nation's energy and transportation systems. Rather than detaching solar from questions of land use and transportation, they viewed renewable energy in an integrative fashion. The Oregon-based artist Diane Schatz produced a series of posters titled *Visions of Ecotopia*, which depicted soft-path, alternative futures for urban, rural, and suburban communities (fig. 10.2). Distributed by *RAIN* (an appropriate technology magazine), her posters moved beyond the individualist frame to promote a collective vision of environmental hope. In the posters solar does not appear as a techno-fix, but instead as a vital, renewing force in a redesigned landscape. Schatz expanded the scale of change to envision not just solar pioneers tinkering in their individual homes but entire communities striving for long-term sustainability. Rooftop gardens and community farming, wind genera-

FIGURE 10.2. "Urban Ecotopia." Poster by Diane Schatz, 1976. Courtesy of Tom Bender.

tors and solar equipment, widely-available public transportation and generously-proportioned pedestrian and bike paths all suggested new models of development and community design. By transcending the personal scale, Schatz offered readers of *RAIN* a broader, community-oriented glimpse into a soft-path future.[9]

Likewise, Denis Hayes and other energy activists hoped to move beyond the individualist frame to outline the broader meanings and political implications of renewable energy. For Hayes, the point of the alternative energy movement was not simply to promote the allure of the sun, but rather to rethink the dominant cultural assumptions about progress and the growth imperative. In his 1977 *Rays of Hope*, Hayes challenged the ideas about time, history, and the future that underlay projections of energy use—projections that always pointed upward toward ever-increasing consumption. According to Hayes, despite their facade of objectivity and reliance upon empirical data, corporate and government projections politicized history: they marshaled statistics from the past to extrapolate into the future without imagining any alternatives to the present energy system. Their use of data reinforced the

159

status quo and rejected alternative visions of the future. These studies, he argued, "make no attempt to grapple with the question 'What can be?' They ask only 'Where do we seem to be heading?' Projections are judgments made today about tomorrow using data generated yesterday." By presenting increased energy use as an inexorable force, Hayes wrote, "the past is widely presumed to be prologue to the future." This forecasting thus denied human agency and obscured the prospect of people and their societies designing different energy systems. These quantitative projections, Hayes claimed, also falsely equated rising energy consumption with "human well-being." Like other alternative energy activists, Hayes instead emphasized "the qualitative dimensions of life" and argued that a different energy system would ensure a more hopeful and sustainable future.[10]

When filtered through the mass media, the ideas of the alternative energy movement would appear overly romantic, disconnected from reality in their ethereal glimpse of a better future. Yet proponents of energy conservation and solar power framed the issue differently: rather than implying that solar would immediately and almost miraculously replace fossil fuels and other nonrenewables, they viewed the contemporary moment as a time of transition. Rather than imagining the future as discontinuous, a sudden break from the present marked by an abrupt, absolute transformation in the energy system, they described the late 1970s as a bridge from one energy regime to another. At the close of *Rays of Hope*, Hayes described the present in these terms: as a moment of passage, a period in which fossil fuels would still be needed as links to the future. "Most energy policy is still framed as though it were addressing a problem that our grandchildren will inherit," he observed. "But the energy crisis is *our* crisis. Oil and natural gas are our principal means of bridging today and tomorrow, and we are burning our bridges."[11]

Energy activists also sought to build new coalitions, to form multiracial, cross-class alliances that would frame renewable energy as simultaneously an economic, environmental, and social justice issue. In New York City, for example, an alternative energy architect provided assistance to a group of low-income, predominantly Puerto Rican and African American residents who formed a tenant-owner cooperative. After taking possession of an abandoned tenement building on the Lower East Side, the group installed solar collectors and a wind turbine on the roof. Their efforts were celebrated by the photographer Jon Naar, who traveled across the United States during the mid-1970s, documenting myriad examples of solar energy projects. Besides the familiar stories of individual tinkerers and solar pioneers, Naar also stressed the collective meanings and empowering possibilities of renewable energy. With the Empire State Building and other prominent skyscrapers in the background, Naar depicted a group of residents on the roof, working together to gain more control over their lives, lower their energy costs, and strive for environmental sustainability (fig. 10.3). This project, together with other examples in the Bronx and Brooklyn, all sought to

FIGURE 10.3. Building at 519 East 11th Street, New York, with tenants. Photograph by Jon Naar, 1976. Copyright Jon Naar, jon@jonnaar.com.

demonstrate that community control over energy resources linked environmentalism to issues of social justice, including entrenched poverty, abandoned buildings, rising rents, and soaring energy costs.[12]

Similarly, Sun Day's board of directors included a diverse range of activists and politicians: environmentalists such as David Brower of Friends of the Earth and Michael McCloskey of the Sierra Club; labor leaders such as Douglas Fraser of the United Auto Workers and William Winpisinger of the International Machinists Union; and scientists, consumer advocates, and elected officials, including Tom Bradley, the first African American mayor of Los Angeles. Through these coalition-building efforts, event organizers hoped to show that solar energy appealed to a broader constituency than the primarily white, affluent membership of mainstream environmental groups.[13]

In particular, Sun Day intersected with one of the most innovative but least reported and remembered dimensions of 1970s environmentalism: the effort to link labor and environmental concerns into a common struggle committed to economic and environmental sustainability. Along with progressive labor leaders, alternative energy advocates formed a short-lived but politically astute group called Environmentalists for Full Employment (EFFE). EFFE leaders believed that mainstream framings of environmental and antinuclear movements too often depicted these struggles as being opposed to the concerns of labor and working people. To counteract this vision of labor and environmentalists "at each other's throats," EFFE produced studies demonstrating that investment in renewable energy would create many more jobs than would the capital-intensive hard path of nuclear and fossil fuels. As Douglas Fraser, the president of the United Auto Workers and an EFFE supporter, explained, "Today, when our nation needs both safe energy and good jobs, solar power can provide both. . . . Properly planned for and developed, such an industry would stimulate economic growth while helping to preserve the environment and safeguard public health."[14]

These visionary efforts to promote green jobs and link environmentalism to the labor movement proved tenuous, and suffered from broader structural changes in the economic and political realms. While EFFE called for large-scale environmental and labor law reform, these ideas stand as a testament to what the historian Jefferson Cowie terms "the New Deal that never happened." Together with other progressive labor leaders, Fraser sought to ally with environmentalists, civil rights activists, and other groups to mount an "anti-corporate offensive" and overcome the "despair at the failure of the political process to respond to the needs of the American people." Their hopeful collective vision, though, failed to counteract the rightward antilabor shift in American politics and the increasing faith in free-market economics as the all-purpose solution to the nation's problems. The neoliberal offensive, which gathered steam during the late 1970s, rejected collective action and government intervention in favor of upholding the market as the preferred means to resolve social and economic troubles. As Cowie concludes, "It all ended . . . [as] something closer to a requiem for a collective economic vision for the American people." Sun Day represents a forgotten moment in a longer history of environmental hope, an event misconstrued by the media as solar-power therapy rather than part of a broader coalition-building effort to improve the nation's energy and environmental future.[15]

As Sun Day approached, Hayes and other solar activists hoped that it would become "an even more politically significant event than was its forebear of 8 years ago, Earth Day." Sun Day leaders wanted to show that alternative energy was more than a random assemblage of tool tinkerers and countercultural individualists reveling in their off-the-grid status. Instead, they wanted to present their cause as "a broad-based movement" seeking "to push politically for rapid development of solar energy." Sun Day marked the moment when alternative energy advocates sought to constitute themselves as a movement—to appear not simply as isolated tinkerers or romantic visionaries, but rather as a collective force seeking to redirect the nation's energy path.[16]

That evening, all three networks devoted several minutes of news coverage to various Sun Day events held around the nation. At Cadillac Mountain in Maine, reputed to be the place in the United States where the sunrise first occurs, several hundred people hiked to the summit to hail the sun's rays. At United Nations Plaza in New York City, the actor Robert Redford spoke about the sun, while in Washington, DC, thousands "spent a day reveling *en masse* . . . at the Washington Monument, which acted as a gigantic sundial." Meanwhile, President Jimmy Carter visited Golden, Colorado, home to the Department of Energy's Solar Energy Research Institute, where he waxed enthusiastic about solar power's potential. "Nobody can embargo sunlight," Carter proclaimed. "No cartel controls the sun. Its energy will not run out. It will not pollute our air or poison our waters. It is free from stench and smog. The sun's power needs only to be collected, stored and used."[17]

National TV coverage of Sun Day emphasized movie star participation in the event, with popular programs such as *The Today Show* and

Good Morning America featuring interviews with Robert Redford and Jane Fonda. "The emphasis on celebrities was regrettable," one Sun Day organizer commented, "since it both implicitly and explicitly linked the concept of solar with the affluent middle class." In a prime example, she continued, "the Jane Fonda interview ended with a close-up shot of the actress' solar powered intercom set in her Los Angeles garden." As in the profiles of Harry Thomason, Jules Bergman, and other suburbanites, solar once again appeared as a form of enlightened consumerism. From the detached homes of middle-class suburbia to the exclusive domains of Hollywood stars, the solar future would be constructed by individual property owners, not by a fundamental change to the nation's energy path.[18]

While all the networks and popular news magazines covered Sun Day events, the overall tone of these reports differed strikingly from that of previous stories, even at times becoming openly hostile. Sun Day, its organizers hoped, provided an opportunity to stage a media event and reach a broader public, to move beyond the individualist frame and allow audiences to glimpse the collective meanings of this energy source. Yet TV and other media outlets seemed unable to accept this broader vision. Indeed, Sun Day coverage often mocked the participants, as reporters adopted a bemused tone and ironic frame to dismiss the event and question the practical implications of alternative energy.

All three TV networks used the weather—that ultimate marker of media temporality, with its focus on immediate conditions—to undermine the political significance of Sun Day. Rain and overcast skies dominated in many places, and TV correspondents took perverse pleasure in noting how the sun refused to cooperate. The day's weather, they implied, cast judgment on the grand hopes of solar advocates. All the news broadcasts noted that Carter encountered clouds and rain in normally sunny Colorado, conditions "that made it tough to assemble the solar models" intended for display. While crowds basked in the sun listening to folk music on the Mall in Washington, flash floods besieged New Orleans, and even the Cadillac Mountain hikers, hoping for a resplendent sunrise, instead witnessed a cloudy dawn.[19]

The most blatantly negative coverage appeared on NBC, where the news broadcast skewered the participants for their misguided dreams and malfunctioning devices. After moving from the National Mall, filled with sunbathers using "the sun for personal purposes," to UN Plaza, where participants "mostly listened to music and watched the sun," the segment then focused on Sun Day events in Atlanta. "But the star attraction wasn't out today," the correspondent announced. "In Atlanta, it was cold and gray." The absence of sunlight on Sun Day, of all days, portended the utter failure of solar activism—or so claimed NBC. Every image and comment from the Atlanta report demonstrated the folly of Sun Day. Some activists hawked pro-solar T-shirts, but it was "too cold to wear" them; a realtor, clutching a book about solar building design, tried to sell people on solar heating, but no one was interested in his spiel. Georgia Tech students "wanted to show off their latest inventions," including solar-powered cookers. "But the hot dogs stayed cold and the enthusiasm lukewarm. They hauled out a solar-powered steam generator, but couldn't get it to work." The correspondent closed by conflating gloomy weather with political failure: "The weather forecast here—rain today, more tomorrow. Sun Day in Atlanta didn't have a prayer." Back in the studio, NBC anchor David Brinkley recapped the day's events by dismissing the entire enterprise of alternative energy as nothing more than romantic delusion. "The ancient Egyptians worshipped the sun as the source of all life, which it is, and called it Ra," he observed. "And the romantic, poetic affection for its beauty and warmth are pleasing and easy. But actually to put the sun's heat to work in useful ways to fit mankind's needs raises questions of costs and practicality."[20]

While TV news and popular magazines had widely praised Earth Day for mobilizing public concern about the environmental crisis, Sun Day would be mocked for betraying the same public desire, for seeking to constitute alternative energy as a viable social movement. Individual tinkerers had been treated with respect, but energy activists were ridiculed. Bringing solar power to individual homes seemed worthwhile, an effort the media endorsed. Yet when energy activists entered the public realm and sought to envision an alternative energy future, the media suddenly cast them as romantic misfits of little consequence. "As if drawn by some pagan rite," *Newsweek* scoffed, "uncountable thousands of Americans assembled at dawn one morning last week on the nation's peaks, parks and promontories to greet the sunrise. They chanted, and waved pale arms aloft, and sang and drank, and smoked organic compounds. And they called it Sun Day, though the sun did not always appear." Even though Sun Day organizers believed the demonstrations represented "the emergence of a new force—a people's coalition to spur the development and use of solar energy," *Newsweek* reined in such grand political hopes. "That may be making a little much of folks' inclination to take a pleasant morning off from work and gape at movie stars," such as "Robert Redford . . . and Jane Fonda," who "did their bits in various places," the magazine chided. Sunbathing, stargazing, and pot smoking, the Sun Day participants did not form a serious political movement, but rather engaged in empty symbolism. Frolicking in the sun, they embraced "the mythic appeal" of solar, an energy source that failed to offer "general practicality."[21]

Not surprisingly, the conservative *National Review* relished the opportunity to deride Sun Day: "They rose early across the country on Sun Day . . . the back-to-nature crowd, brown-rice-and-carrot devotees, zen-joggers—all up to worship the sun as it hasn't been worshipped since the Aztecs. It was undoubtedly one of the most impressive tribal gatherings since Earth Day in April 1970." The article then shifted from contempt to concern, worrying that the event might join its predecessor in leading to major policy changes. "No doubt, most of us . . . rolled over and went back to sleep, dismissing it all as just another freaky reunion of the aging Sixties Generation," the article continued. "And it may have been. But many of us also felt that way about Earth Day, not quite fully realizing that the ethos it symbolized would culminate in such things as the Environmental Protection Agency."[22]

The *National Review* need not have worried. Even though the "ethos" of mainstream environmentalism had been enshrined, at least in a limited way, in the EPA and the creation of the environmental regulatory state following Earth Day, the same could not be said of Sun Day. The ethos of energy activists—the extensive timescale of the soft path, the link to futurity, the collective vision, and the involvement of labor and other social causes—all these would be ignored or denigrated by the media.

While the antireactor movement could align itself with the sense of immediate crisis and fear of technological collapse popularized by *The China Syndrome* and Three Mile Island, alternative energy advocates could not benefit from a similarly punctuated experience of time. Lacking the spectacle of potential meltdown, the media framed solar energy in the following ways: as a benign, even laudable pursuit carried out by isolated individuals; or as an idle dream of starry-eyed protesters who did not belong in the public realm, who deserved to be ignored and cast aside by policy makers.

The mainstream media, though, was not completely at fault for presenting solar energy as a form of do-it-yourself activism. Indeed, many alternative energy advocates viewed solar in a techno-determinist fashion, believing that the small-is-beautiful utopia would emerge simply from individuals making "choices about the right technologies," rather than from collective struggles over power and politics. This mode of "ecological individualism" and "technological optimism" often became, much like the "Don't Be Fuelish" campaign, a depoliticized effort to demonstrate personal virtue. Not long after Sun Day, the political theorist Langdon Winner critiqued the "better mousetrap theory" of social change that guided these efforts: the idea that individual use of solar technologies would "serve as a beacon to the world, a demonstration model to inspire emulation." "If enough folks built for renewable energy, so it was assumed, there would be no need for nuclear power plants," Winner explained. "People would, in effect, vote on the shape of the future through their consumer/builder choices. . . . Radical social change would catch on like . . . some other popular consumer item."[23]

The techno-determinism of the alternative energy movement can also be seen in popular efforts to frame antinuclear/pro-solar positions as a simple dichotomy, a choice between radioactivity and sustainability, the ominous cooling towers versus the luminous rays of the sun. In September 1979, a few months after Three Mile Island, the Musicians United for Safe Energy (MUSE) sponsored a series of concerts that featured James Taylor, Jackson Browne, Bruce Springsteen, and other musical celebrities, and which led to a *No Nukes* triple live album and concert film. The album art portrayed the sun as a glowing, beneficent body, completely detached from human society and earthly problems. "The three albums come wrapped in a splendid copper-colored art-deco sun, radiating wavy lines of solar force," one reviewer noted. Much like protest imagery that pictured the sun along with the slogan "We don't need nuclear power," MUSE iconography promoted the better mousetrap theory of social change, depicting the copper-colored sun radiating beams of hope, a technical fix to the nation's energy and environmental crises.[24]

Rather than offering a radical vision of social change or challenging the expansionary ethos of consumer capitalism, popular depictions of solar energy ultimately served to blunt the critical edge of the environmental and alternative energy movements. Soon after Three Mile Island, TV news broadcasts would note that solar was "popular with just about everybody." Yet this notion of unified support for solar obscured the critical questions asked by many environmental activists about the escalating risks of fossil fuel dependency and the ecological limits to growth. Solar appeared as a small, sustainable subset of the nation's total energy picture—not as a vibrant image of environmental hope, nor as part of a broader vision to reorient the energy future.[25]

While Hayes hoped that Sun Day would be even more influential than Earth Day, the event's impact on national energy policy would be remarkably circumscribed. Perhaps the most tangible expression of Sun-Day–induced change would be found, the following year, on the White House roof. On Sun Day, as President Carter voiced his commitment to solar energy, his administration announced its plan to install a solar water heating system at the White House. Solar panels appeared atop the West Wing the next year. While they have often been hailed as an effort to green the White House, energy activists viewed them instead as a form of presidential greenwashing. Despite the image of sustainability promoted by the panels, they argued, the president appeared as tinkerer-in-chief, a technocrat devoid of a larger social vision, a hard-path aficionado who cast soft-path solutions primarily in the narrow mode of ecological individualism. While Sun Day sought to reframe the emotional politics of environmentalism, to inspire with images of hope, Carter would soon retreat to the familiar discourse of environmental guilt.[26]

ELEVEN

Carter's Crisis and the Road Not Taken

On June 20, 1979, Jimmy Carter held a press conference in an unusual place: on the roof of the White House. Standing in front of thirty-two recently installed solar panels, Carter tried to imagine the future significance of this White House effort to promote solar energy (fig. 11.1). He assumed that the panels themselves would be there for decades to come, and hoped they signaled the promise of a more sustainable energy future. "In the year 2000, the solar water heater behind me . . . will still be supplying cheap, efficient energy," he announced. "A generation from now, this solar heater can either be a curiosity, a museum piece, an example of a road not taken, or it can be just a small part of one of the greatest, most exciting adventures ever undertaken by the American people."[1]

Seven years later, the solar panels were removed. Many environmentalists—from 1986 until today—have considered their removal emblematic of the Ronald Reagan administration's effort to dismantle environmental regulations. The panels eventually ended up at Unity College in Maine, where they were used to heat water for the cafeteria. While environmentalists often described the panels as a sign of Carter's dashed hopes and Reagan's misguided policies, the panels themselves remained, for the most part, hidden from public view.[2]

In 2010, though, they became more visible. That spring, two Swiss filmmakers released a documentary titled *A Road Not Taken* that followed the solar panels from

FIGURE 11.1. President Carter at White House solar panel dedication, 1979. Copyright Bettmann / Corbis / AP Images.

the White House to Unity College and beyond. The film presents the White House experiment in alternative energy as a road not taken for a nation that has become, in successive decades, even more dependent upon nonrenewables. Later that year, Bill McKibben, a bestselling environmental author and founder of the climate change activist group 350.org, captured media attention by joining a few Unity College students on a road trip to the White House. Hoping to convince Barack Obama to reinstall the panels, they also protested his administration's poor record on energy and climate change. "Obama has drawn much of the blame for the failure of the climate legislation, which he didn't push aggressively; this is a chance to make at least symbolic amends," McKibben observed. "Clearly, a solar panel on the White House roof won't solve climate change—and we'd rather have strong presidential leadership on energy transformation. But given the political scene, this may be as good as we'll get for the moment." Although Obama refused to put the vintage panels on the roof, his administration claimed that new solar panels would be installed in 2011. Almost three years after this pledge, the panels finally began to grace the White House roof.[3]

It is important to recognize, though, that the sense of hope and possibility attached to Carter's solar panels accrued only over time, representing, according to many contemporary environmentalists, both the sustainable path not taken and the senseless myopia of the Reagan administration. Yet, when Carter unveiled the panels on a sunny day in 1979, they attracted relatively little attention and were seen by most environmentalists at the time as an empty gesture, a symbolic distraction from Carter's failures to alter the nation's energy path. While many environmentalists had endorsed Carter in the 1976 election, they became increasingly critical of his energy policies. "As the administration's policy unfolded, environmentalists were most disappointed by the imbalance between nuclear and solar options," the historian Samuel Hays explains. "Budget proposals to Congress in 1978 called not for continued increases in solar-energy funding but for reductions." During the second half of Carter's term, environmentalists "found his energy policy to be the most disappointing aspect of his environmental record." Indeed, less than a month after the solar panels appeared, leaders from many environmental groups—mainstream and radical alike—condemned Carter's policies for reinforcing the status quo.[4]

Today, photographs of Carter standing in front of solar panels seem captivating, even inspiring, suggesting that this visionary president sought to lead Americans toward a sustainable energy future. For all their retrospective allure, though, the images need to be viewed more critically, to be placed within their historical context to reveal the limits to Carter's environmentalism. Carter's problematic responses to the energy crisis—including, most notably, the televised "Crisis of Confidence" speech—too often emphasized individual culpability for wasteful consumer habits but failed to outline a broader social ethic or envision structural changes to the nation's energy systems. The popular memory of Carter as an environmental president obscures the alternative energy future imagined by environmentalists, and wrongly conflates the discourse of individual responsibility with the political vision of American environmentalism. This perspective—widely circulated by the media, routinely deployed by conservative politicians, and often accepted by professional historians—silences the voices of Carter's critics and assumes that environmental activists focused solely on small-scale personal decisions about energy use but failed to engage in larger power struggles over American energy policy. Even as Carter joined environmentalists in trying to focus public attention on the energy crisis, his plans too often gave short shrift to the soft path, choosing instead to take the road more traveled by.

Carter agreed with environmentalists that the oil shocks of 1973–74 should be viewed not as a short-term problem of supply caused by OPEC, but rather as a long-term crisis that the nation needed to confront. Almost from the moment he took office, Carter used TV as a means to mobilize public concern around energy issues, delivering speeches that blended crisis rhetoric with appeals to common sacrifice. In February 1977, just two weeks after his inauguration, Carter gave the first in a series of televised energy addresses. Sitting by the fireplace and looking a bit like Mister Rogers, the popular star of a children's TV show, Carter donned a cardigan sweater to signify the need to turn down the thermostat and to demonstrate that even the president was taking personal responsibility for energy use.[5]

Afterwards, CBS veteran correspondent Eric Sevareid chuckled as he asked anchor Barry Serafin, "Barry, did you ever see a president sitting before the whole country in a sweater before?" Answering his own question, Sevareid continued, "I never did." Sevareid noted that the curious cardigan reflected both the "somewhat chillier" temperature in the White House and the "more informal" tone the president hoped to strike. Carter, another correspondent explained, wanted to go "directly to the people." Even as Carter relied on TV to deliver his message, he hoped that his approach bypassed media filters to communicate with the citizenry in an honest, straightforward fashion. He used visual symbols such as the cardigan sweater and the fireplace to convey a folksy image, to present himself as a man of the people, someone who wanted to participate in a shared project of energy conservation to safeguard the future.[6]

In his initial energy speeches, Carter urged Americans to keep their attention directed toward a crisis that many thought had long since passed. While media temporality often blocked understanding of the long-term implications of American energy use, Carter stressed that "the energy shortage is permanent" and that the United States needed "to plan for the future." Carter's speeches not only tried to maintain a sense of urgency regarding energy issues, but also gestured toward a broader collective vision of citizenship in an age of limits. He tried to explain to viewing audiences how their personal energy decisions intersected with larger problems facing the nation. For Carter, the energy crisis represented "the moral equivalent of war" and demanded a concerted, collective response by the American people. Like environmentalists who sought to move beyond the narrow vision of citizens as consumers that had been promoted by gas line imagery, Carter's rhetoric expanded the notion of citizenship to include responsibility for the common good and for the future. He even suggested that particular forms of consumption would need to be scaled back—or at least made more costly. "Those citizens who insist on driving large, unnecessarily powerful cars must expect to pay more for that luxury," Carter claimed.[7]

Environmentalists endorsed some elements of Carter's rhetoric and policies. They agreed with the president that the atomized, privatized vision of citizenship fostered by gas line imagery and consumer culture worked to separate Americans from one another and to deny their dependence upon the earth's finite resources. Environmentalists also supported measures to increase automobile fuel efficiency and to encourage conservation in homes and workplaces. While they worried about the overconsumption of resources, they tended to view this problem in a systemic manner; they emphasized how large-scale, structural decisions made by corporations and governments ultimately determined the nation's increasing levels of energy use and the continued reliance upon nonrenewable sources of power.

Despite the modest changes enacted by Carter's energy reforms, many environmentalists believed that the president failed to take seriously the soft-path vision of conservation and renewable energy. The trouble with Carter, they began to realize, was that his moralistic rhetoric obscured power relations and his meager soft-path gestures marginalized their vision of an alternative energy future. These problems became most apparent in the summer of 1979, only a few months after *The China Syndrome* and Three Mile Island had popularized fear of nuclear power, and a few weeks after solar panels had appeared on the White House roof. During that summer, as gas lines returned with a vengeance, the president went on TV to sermonize about energy.

Delivered on July 15, 1979, the "Crisis of Confidence" speech—often referred to as the "malaise" speech—followed a ten-day retreat at Camp David where the president, his advisers, and 135 consultants pondered the energy crisis and the larger "crisis of the American spirit." "It was a perfectly executed political drama with an Old Testament twist," one commentator observed. At the end of the retreat, "the President, described in the press as an American Moses, descended from his Catoctin mountaintop armed with a new energy policy."[8]

Yet rather than becoming a prophetic figure leading his people to a new covenant, Carter would largely be remembered as someone who, especially in this address, scolded Americans and failed to inspire hope in a time of crisis and uncertainty. "By sermonizing . . . ," the historian Sean Wilentz argues, "Carter appeared to be abdicating his role as leader and blaming the people . . . for their own afflictions." Carter's pessimistic vision would be refuted by the right and skillfully exploited by the Reagan campaign the following year. Denying the reality of ecological limits, Reagan offered up a simplistic, reassuring rhetoric of hope and confidence to dispense with Carter's specter of malaise.[9]

The immediate reaction to the speech was in fact more positive than is commonly noted: Carter enjoyed a brief spike in approval ratings, and many Americans soon wrote letters to the president, praising his effort to speak so candidly about the nation's "crisis of confidence" and promising to take personal action—such as "riding a bike or moped to work" or "turning down the thermostat"—to reduce the nation's "reliance upon foreign oil." Seeking to rescue Carter from the condescension of posterity and hoping to resuscitate the speech from the amnesia produced by the popular media and history textbooks, the historian Kevin Mattson argues that "we *mis*remember the speech today. Most of us know it as Carter's 'malaise' speech, even though Carter never used that word." Mattson emphasizes subsequent events—especially Carter's dramatic reorganization of his cabinet a few days later—as shaping the overwhelmingly negative view of the speech that soon formed. This forgetful legacy, he believes, continues to deny Carter's ideas their rightful place in cultural memory and fails to acknowledge their thwarted potential to help the nation respond to the energy crisis.[10]

The trouble with the popular memory of Carter is that it also enshrines the false notion that his denunciation of consumerism and appeal to individual responsibility somehow represented the vision held by environmentalists and reflected their diagnosis of the energy crisis. Indeed, what gets ignored in all these accounts is the alternative energy future envisioned by environmentalists as well as their broader critiques of energy policy. Carter's address, and the media-produced memory formed in its wake, worked to delimit popular understandings of the energy crisis and to derail the more far-reaching ideas and hopes of environmentalists. For this reason, we need to attend to the speech's silences and absences: its effort to marshal a civic vision that denied power relations; its condemnation of consumerism, which concealed larger structures of power; its inability to imagine an ecologically sustainable, more democratic path out of the abyss.[11]

Carter delivered the speech soon after gas line imagery had returned to TV and other media sources, as shortages triggered by the Iranian Revolution had once again led to long lines at the pumps. The pictures of the gas lines resembled the iconic images from the 1973–74 crisis in their focus on the inconvenience experienced by consumers. Carter, though, sought to transcend the visual media's emphasis on the immediate crisis of gas lines and to develop instead a broader cultural diagnosis of the crisis of the American spirit. "It's clear that the true problems of our nation are much deeper—deeper than gasoline lines or energy shortages," he argued. Rather than focusing on media-induced notions of temporality and momentary flare-ups of crisis, rather than limiting his analysis to the anger felt by drivers lining up for scarce supplies of fuel, Carter wanted to explain the long-term causes of the nation's rising cynicism and collective feelings of disengagement. He wanted to move beyond the headlines to assess the psychic state of the American people.[12]

Carter and his speechwriters drew on the historian Christopher Lasch's *The Culture of Narcissism* (1979) to explain how consumer culture had eroded public life and fostered a narcissistic society more concerned with the acquisition of material goods than with the creation of a sustainable future. "In a nation that was proud of hard work, strong families, close-knit communities, and our faith in God, too many of us now tend to worship self-indulgence and consumption," Carter argued. "But we've discovered that owning things and consuming things does not satisfy our longing for meaning." Ignoring class divisions and racial inequities within US society, Carter deployed the familiar language of universal responsibility to blame all Americans for their selfishness and their insatiable pursuit of meaningless things.[13]

After this jeremiad, Carter appealed to the better angels of the American character to encourage audiences to view sacrifice and conservation as pathways to restoring public life and faith in the future. Echoing his earlier speech which had framed the energy crisis as the "moral equivalent of war," Carter again emphasized the patriotic meanings of conservation. He argued that individual energy-saving acts would contribute to the public good and help ensure the nation's collective destiny. "And I'm asking you," he implored, "for your good and for your Nation's security to take no unnecessary trips, to use carpools or public transportation whenever you can, to park your car one extra day per week, to obey the speed limit, and to set your thermostats to save fuel. Every act of energy conservation like this is more than just common sense—I tell you it is an act of patriotism."[14]

Environmentalists, of course, agreed with Carter that Americans should strive to reduce their energy consumption by engaging in everyday acts of conservation. Nevertheless, like a number of left-leaning intellectuals who critiqued the speech, they questioned Carter's failure to consider power relations. While the president railed against consumerism as a moral failing, they instead emphasized corporate structures and government policies that led to unsustainable rates of energy use.

Rather than confronting these issues, Carter instead channeled the message of the Crying Indian to hold individual Americans responsible for the energy crisis. Iron Eyes Cody had actually visited Carter at the White House one year earlier, on the eve of Earth Day 1978, a day that the actor's employer had rechristened as Keep America Beautiful Day. Addressing the president as the "Great White Chief," he "placed a handmade eagle feather headdress on [Carter's] head." Images of Carter and Cody appeared on all three TV networks and in newspapers around the nation (fig. 11.2). Even as environmentalists continued to struggle against the message of the Crying Indian, his presence continued to haunt them. Even as they tried to offer a hopeful, collective vision of the energy future, the twinned discourses of environmental guilt and personal responsibility became central to Carter's "Crisis of Confidence" speech.[15]

While the address was short on policy prescriptions, environmentalists could point to some victories and hopeful signs in the president's recommendations. In particular, the word "nuclear" was completely

FIGURE 11.2. President Carter dons a headdress as he chats with Iron Eyes Cody, 1978. Copyright Bettmann/Corbis.

absent from the speech. Even though nuclear power remained crucial to Carter's energy policy, his speechwriters decided—in the wake
of Three Mile Island and the increasing visibility of the antinuclear
movement—that it would be politically unwise for the president to
mention, let alone extol, this energy source. Having recently dedicated
the solar panels on the White House roof, Carter also proposed "the
creation of this nation's first solar bank, which will help us achieve the
crucial goal of 20 percent of our energy coming from solar power by
the year 2000."[16]

Through these reassuring words, Carter sought to align himself with
the environmental movement. Yet his emphasis on individual culpability, his failure to discuss corporate power and responsibility, and his
soon-to-be-announced energy polices all convinced environmentalists that he was, as one critic put it, "more a Judas than a Moses." Sierra Club president Theodore Snyder argued that "Carter's energy plan
slights energy conservation and renewable energy in favor of [the] hard
path; it simply fails to solve the nation's energy-supply problems and
will result in continuing dependence on OPEC."[17]

The emphasis of Carter's policies, environmentalists and leftist critics claimed, would push the United States further away from the soft,
sustainable path of conservation and renewable energy. "The center-
piece of the Carter energy program is a commitment to the massive
development of synthetic fuels from coal and oil shale," the *Progressive*
magazine explained. "As with nuclear power, the U.S. taxpayer would
bear the risks and large corporations would reap the profits." Carter
sought to expand the production of synthetic fuels through "extracting liquids and gases from coal, oil shale and tar sands." These capital-
intensive extractive processes would generate huge quantities of pollution and require "massive amounts of water" from the arid West and
the mining of millions of tons of additional coal to run synthetic fuel
plants. The plan also called for the "licensing and construction of more
commercial [nuclear] plants." All in all, *Newsweek* noted, "the real focus
of the program is on production."[18]

Environmentalists had, in fact, been shut out of the Camp David retreat. Over the ten-day period, Carter's guest list had included 135 people who consulted on the energy speech, only one of whom—Russell
Peterson of the Audubon Society—hailed from an environmental organization. Environmentalists felt so marginalized that they decided to
take action: representatives of the Sierra Club, Friends of the Earth, and
other leading groups "hand-carried a letter" to the president at Camp
David, "warning that synfuels aggravate air pollution, release carcino-

gens and could put a serious drain on water resources in the arid West." "We are concerned that you are on the brink of making a disastrous and irreparable mistake in your choice of energy strategies," the environmentalists concluded.[19]

Just as environmentalists challenged Keep America Beautiful for its propagandistic focus on personal responsibility, they now pointed to the limitations of Carter's speech by situating individual action within a broader framework. Rather than merely preaching to the American people about their wasteful habits, Carter and the federal government needed to alter energy and transportation policies, to make it more possible for citizens to demonstrate ecological virtue in their daily lives. "But the Carter program will provide neither the resources nor the opportunity for the comprehensive national and local planning needed if we are to achieve a secure and safe energy future," one commentator concluded. "And despite comforting rhetoric about a solar future, the massive open-ended commitment to nuclear and synthetic fuel will leave little capital for safer, more effective alternatives."[20]

Christopher Lasch himself joined environmentalists in lambasting Carter's diagnosis of the energy crisis. Like other critics of the hard path, Lasch believed that Carter's plan emphasized systems of power production that would only benefit "the very companies" most responsible for causing the "current critical pass." Even though Carter sought to extend the time frame of crisis and move beyond media temporality, his policies demonstrated environmental myopia. "A centralized policy relying heavily on nuclear power and other synthetic fuels," Lasch continued, "is objectionable not only because it perpetuates the morally indefensible concentration of power and wealth . . . but because it mortgages the future to the immediate interests of the present generation." Carter may have sermonized about the immorality of consumer culture, but his plan sacrificed the future. Lasch concluded, "This is the ultimate indictment of the 'culture of narcissism'—not that it is self-indulgent and self-absorbed but because it is criminally indifferent to the welfare of the next generation and the generation after that."[21]

Tony Auth, a *Philadelphia Inquirer* editorial cartoonist whose work was frequently reprinted in the *Progressive* and other left-leaning magazines, visualized this critique of Carter's disregard for the future. Soon after Three Mile Island, Auth produced several cartoons that deployed the smiley-face symbol—one of the most iconic symbols of 1970s America—to condemn nuclear energy, turning this emblem of cheer into a disturbing sign of the industry's effort to mask radioactive danger. In response to Carter's energy address, Auth created a densely-

FIGURE 11.3. Editorial cartoon by Tony Auth, 1979. AUTH © The Philadelphia Inquirer. Reprinted with permission of UNIVERSAL UCLICK. All rights reserved.

packed cartoon that mobilized many elements of the environmentalist critique of the president (fig. 11.3). Auth placed a smiley face near the top of a cooling tower and also portrayed other markers of the hard path: polluting smokestacks along with earthmoving equipment excavating the ground, presumably in an attempt to find raw materials for synthetic fuel production. In the foreground, Carter has just removed a gas mask, that iconic symbol of pollution and universal vulnerability. The president leans down to speak to a flower and says, "We're all going to have to make sacrifices . . . especially you, Mother Earth." Reminiscent of the Daisy Girl, Auth's cartoon depicts the flower as an emblem of nature and futurity. Yet while Lyndon Johnson had presented himself as a defender of life and promoter of environmental citizenship, Carter appears as someone with little concern for the environmental future. A visual counterpoint to the popular memory of Carter, this cartoon reflects how many environmentalists viewed him at the time.[22]

While Carter is often remembered for the "malaise" speech, the Carter Doctrine, first articulated in his 1980 State of the Union address, exerted a much greater impact on American policy and dominant ideas of fossil fuel futurity. In response to the ongoing hostage crisis in Iran and rising oil prices, Carter announced that the US would maintain its access to Persian Gulf oil "by any means necessary, including military force." Rather than imagining a post-petroleum society or promoting a soft-path vision of the future, the Carter Doctrine justified the expansion of military operations in the Persian Gulf region and suggested that the Pentagon could ensure the long-term availability of cheap oil. This militarized approach encouraged an increasing dependency on imported fossil fuel and established the policy framework for the Persian Gulf War of 1991 and the Iraq War that began in 2003. Together, the hard path and energy securitization promoted the fantasy of perpetual petroleum, and led to blood being spilled for oil.[23]

Around the same time that Carter delivered his "Crisis of Confidence" speech, the Ad Council launched a new energy conservation campaign. Produced on behalf of the Alliance to Save Energy, a nonpartisan group that encouraged individuals to reduce their personal energy consumption, the campaign sought to convince Americans that the nation's unsustainable patterns of energy use ultimately threatened future generations. The actor Gregory Peck starred in TV spots and print ads laden with post-frontier imagery. Outfitted in western attire, Peck appeared at the edge of a ghost town (fig. 11.4). "This used to be a flourishing, gold rich town," he observed. "People who lived here never thought it would end up like this, but . . . the gold ran out." Strolling through abandoned streets and walking into dilapidated buildings, Peck continued, "Today, we run a risk of losing something more precious than gold, our country's energy. We waste a shameful amount of it." Mapping frontier ruins onto contemporary energy gluttony, the campaign asked viewers to change their wasteful ways. "For our children and *their* children . . . ," one tagline announced, "let's save energy now!"[24]

While environmentalists had hoped to use the visual media to help Americans imagine an entirely different energy future, Carter and the Ad Council left audiences with a fear of scarcity that could only be addressed by individual action. Both Carter and the Ad Council appealed to the idea of governing at a distance, of citizens engaging in personal action to reduce the nation's energy consumption. These appeals, once again, joined the Crying Indian and previous energy PSAs in claiming that people started this crisis, and that people could stop it.

Through televised addresses and PSAs, Carter and the Ad Council engaged with the emotional politics of the era and sought to overcome the anger and alienation felt by many Americans. One Carter adviser even hoped that the "Crisis of Confidence" speech would resemble Howard Beale's "mad as hell" tirade but channel public response into

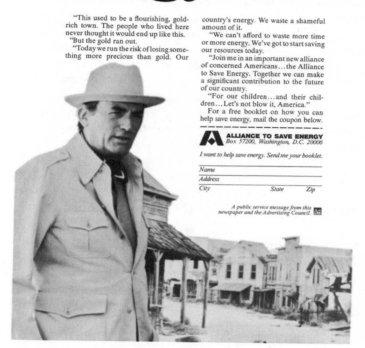

For our children and their children...

Let's save energy now!

"This used to be a flourishing, gold-rich town. The people who lived here never thought it would end up like this. "But the gold ran out.

"Today we run the risk of losing something more precious than gold. Our country's energy. We waste a shameful amount of it.

"We can't afford to waste more time or more energy. We've got to start saving our resources today.

"Join me in an important new alliance of concerned Americans...the Alliance to Save Energy. Together we can make a significant contribution to the future of our country.

"For our children...and their children...Let's not blow it, America."

For a free booklet on how you can help save energy, mail the coupon below.

ALLIANCE TO SAVE ENERGY
Box 57200, Washington, D.C. 20006

I want to help save energy. Send me your booklet.

Name

Address

City   State   Zip

A public service message from this newspaper and the Advertising Council.

FIGURE 11.4. "Let's save energy now!" Advertising Council / Alliance to Save Energy advertisement featuring Gregory Peck, 1979. Courtesy of the Ad Council Archives, University of Illinois, record series 13/2/207.

positive solutions to the energy and cultural crises. Ultimately, though, the frustration felt by drivers in gas lines and the outrage expressed by Beale's cinematic rant translated not into a sustainable energy future, but rather into widespread feelings of anger and weakness, of a nation held hostage by "oil sheiks" and other external threats. Republican politicians and free-market activists tapped into this anger to organize anti-tax initiatives and other campaigns to limit the scope of government regulatory agencies and relief programs. Indeed, California's in-

181

fluential tax revolt, which culminated in the passage of Proposition 13 in 1978, adopted Beale's "mad as hell" line as its slogan. Public feelings of anger and frustration coalesced into what the political scientist Evan McKenzie terms the "ideology of hostile privatism" underlying the triumph of neoliberal values and policies: market deregulation, low taxes, and the upward redistribution of wealth.[25]

While some viewers accepted Peck's post-frontier vision of limits and Carter's sermonizing about the energy crisis, Ronald Reagan invoked the popular myth of frontier abundance to repudiate environmental guilt, and cast Carter as an emblem of "pessimism, fear, and limits." In 1980 Reagan's presidential campaign mobilized frontier imagery to portray him as a rugged, self-reliant Westerner (fig. 11.5). Publicity photographs depicted him wearing a cowboy hat and astride a horse, a confident man of action, a Western hero who could overcome the age of limits. Drawing on familiar motifs from film and TV Westerns, this imagery showed Reagan in control of his surroundings, a man who could break horses and chop wood, a man whose body conveyed strength and virility, a man who could lead the nation out of the despair and malaise of the 1970s. The image of Reagan as cowboy linked masculinity to the control of nature, to the ability of the

FIGURE 11.5. Ronald Reagan presidential campaign button, 1980. Courtesy of Ronald Reagan Library.

president—and, by extension, the nation—to subdue the frontier and exploit its abundant resources.[26]

In contrast to this image of Reagan as powerful and confident, Carter was often portrayed by the media as weak and ineffective—as a man who even became terrorized by a bunny. Just a month after the "Crisis of Confidence" speech, the news media seized upon a seemingly improbable event: a "killer rabbit" had attacked Carter while he was on a fishing trip in Georgia. Described by the Associated Press as "hissing menacingly, its teeth flashing, and nostrils flared," the crazed rabbit swam toward Carter's canoe; the president then used a paddle to beat back the bunny. The story eventually made it to the front page of the *Washington Post* and other newspapers and was covered by all three TV networks. One editorial cartoon, titled "Paws," played off the iconic *Jaws* movie poster: A rabbit lurks beneath the placid waters while the president, rather than a female swimmer, floats by in his canoe. The gendered meanings were clear: The rabbit, sensing weakness and vulnerability, prepares to attack the feminized president.[27]

Media images of Carter and Reagan presented a set of binaries that equated masculinity with the control of nature and the restoration of American power and frontier abundance: Carter wearing a cardigan sweater, Reagan wearing a cowboy hat; Carter being attacked by a bunny, Reagan confidently riding a horse; Carter pleading for sacrifice, Reagan projecting a sense of limitless optimism. While Carter espoused the Carter Doctrine, media images portrayed Reagan as more capable of restoring national strength, responding to foreign threats, and regaining access to plentiful supplies of oil.

The 1979 fuel shortage would be followed by the Iran hostage crisis, in which Islamic militants stormed the US embassy in Tehran, took Americans captive, and held them for 444 days. Carter's inability to rescue the hostages deepened the popular perception of presidential weakness and reinforced the sense of American decline, of the nation becoming vulnerable to external powers. The frustrating reprise of fuel shortages, together with the media spectacle of suffering hostages, connoted a crisis of national identity that Reagan promised to resolve through a commanding, masculine mode of leadership. "Adversaries large and small test our will and seek to confound our resolve," Reagan explained, "but we are given weakness when we need strength; vacillation when the times demand firmness." Reagan relied upon frontier imagery to reject the environmentalist concern about limits and to proclaim that masculine prowess could deliver the nation from its enfeebled state.[28]

Environmentalists and alternative energy advocates had tried to inspire hope, to present their movement as offering solutions to the age of limits and to the cynicism and alienation of public culture. Instead, Carter and the Ad Council propagated a privatized vision of energy conservation that constrained popular framings of environmentalism and failed to imagine an alternative energy future. Reagan mobilized hope by denying environmental limits and offering a privatized vision of freedom. With the gas lines gone and with energy prices plummeting in the new decade, Reagan addressed citizens as consumers, as individuals pursuing their private interests, rather than as participants in a collective project. No wonder, then, as Reagan slashed funding for solar research and removed solar panels from the White House roof, Carter would be viewed, retrospectively, as an environmental president.[29]

Environmentalists felt marginalized by Carter in 1979 and mocked by Reagan in 1980, yet the movement, even at this demoralizing moment, could still claim some success in altering energy policy. The public fear of nuclear power, together with the outrageously high cost of nuclear plant construction, led to the cancellation of new plant orders. New fuel economy standards led to modest gains in automobile efficiency, while insulation, weather stripping, and other energy-conserving measures became more common in homes and buildings across the nation. Still, the movement failed to change the nation's transportation policy in any significant way or to convince policymakers to embrace a soft-path vision of the energy future.

Even at this low point, though—even as neoliberal ideology called for the deregulation of industry—environmentalism would still benefit from its ongoing interaction with the popular visual media. Throughout the 1980s and especially in the period leading up to Earth Day 1990, the news media's focus on immediate crises, the increasing involvement of Hollywood stars in the movement, and the resurgence of the discourse of individual responsibility all helped raise the cultural profile of environmentalism. Indeed, the story of green going mainstream is in large part the story of environmental spectacle in a neoliberal age.

PART THREE

Green Goes Mainstream

TWELVE

Environmental Spectacle in a Neoliberal Age

With the 1988 presidential election only two months away, George H. W. Bush boarded a ship in Boston Harbor. He toured the polluted waters and then blamed his opponent, Massachusetts Governor Michael Dukakis, for allowing the storied harbor to become the "dirtiest" in the nation, a veritable "harbor of shame." "It was floating political theater," CBS News explained that evening, "and the plot was simple: show that, while Dukakis talks about cleaning up the environment, Boston Harbor just gets dirtier."[1]

The Bush campaign soon released a TV commercial depicting the harbor as an ecological eyesore. Close-up shots revealed raw sewage pouring out of pipes, chemicals coalescing into putrid patterns of color, and refuse and dead fish washing ashore. At one point a warning sign filled the screen: "Danger / Radiation Hazard / No Swimming." These visuals referenced earlier scenes of environmental crisis—the unsightly spread of pollution and litter decried by Pogo and the Crying Indian, the viscous swirls of color generated by Santa Barbara and other oil spills, and the fear of radiation popularized by various antinuclear movements—to present Boston Harbor as a disturbing sign of environmental degradation. The male narrator gravely commented: "As a candidate, Michael Dukakis called Boston Harbor an open sewer. As governor, he had the opportunity to do something about it, but chose not to. . . .

And Michael Dukakis promises to do for America what he's done for Massachusetts."[2]

Boston Harbor became the most resonant image in Bush's environmental strategy. It became discursive shorthand for his opponent's poor environmental record, the place name alone signifying ecological blight. On the campaign trail, Bush repeatedly evoked the harbor to discredit the Democratic nominee. At one rally, referring to Dukakis's position on a completely different issue, Bush gibed: "If you believe that, he's got some bottled water from Boston Harbor to sell you." Likewise, in a televised debate, Bush responded to a Dukakis comment with the quip: "That answer was about as clear as Boston Harbor."[3]

Neither Bush nor Dukakis boasted an exemplary environmental record, but environmentalists favored the Democrat over the Republican. The League of Conservation Voters, for example, gave Dukakis a "B" on its environmental scorecard, while Bush received only a "D plus." By going negative, by making a spectacle of Boston Harbor, Bush sought not only to attack his opponent's record but also to appropriate the environmental issue, to prove that he would be "the environmental president."[4]

Why, though, did Bush want to be viewed in this way? For eight years he had served as vice president under Ronald Reagan, whose free-market fundamentalism had sought to loosen environmental restrictions and deregulate industry, whose faith in boundlessness had denied ecological limits and repudiated Jimmy Carter's rhetoric of conservation and sacrifice, and whose appointments of James Watt as secretary of the interior and Anne Gorsuch as director of the Environmental Protection Agency had outraged environmentalists. For eight years, Bush had also chaired the Presidential Task Force on Regulatory Relief, which "canvassed business leaders to develop a 'hit list' of regulations to repeal," including policies related to air pollution, pesticides, and toxicity. After eight years of promoting regulatory relief, of rewriting policies to grant corporations more freedom to pollute, Bush now presented himself as an environmental candidate.[5]

It would be easy to dismiss Bush's environmental rhetoric and use of environmental crisis imagery as nothing more than empty campaign politics, a calculated strategy to exploit the public's environmental feelings while masking his own anti-environmental agenda. Indeed, this election-time gambit, many environmentalists charged, constituted a cynical move designed to deceive voters rather than to convey genuine environmental concern. Given Bush's ongoing commitment to deregulation and his disappointing record once in office, environmentalists were certainly not wrong to lambaste his environmental posturing as blatant political opportunism.

Yet Bush's successful effort to frame Boston Harbor as a menacing example of environmental crisis needs to be viewed within a broader field of struggle, a context in which environmental spectacle both challenged and reinforced the logic of free-market ideology. Throughout the 1980s, and especially during the latter part of that decade, the visual media publicized a series of environmental crises—including the ozone hole, the garbage crisis, rain forest destruction, and global warming—that signified inherent ecological limits rather than Reaganite dreams of boundlessness. Media images helped audiences see the ecological consequences of consumer culture. Despite Reagan's notorious federal appointments and wishful denial of limits, the visual media's focus on alarming environmental trends aroused public concern, generated a sense of urgency, and at times even encouraged environmental reform in the midst of a conservative age.

By the summer of 1988, just before Bush cruised through Boston Harbor, the notion of an overarching environmental crisis—a concept first popularized in the time leading up to Earth Day 1970—once again gained increasing purchase in US public culture. The visual media situated various environmental problems within a larger frame to suggest a sense of planetary peril. Bush's environmental claims thus sought to tap into popular environmental concern and to project a public image that tempered the free-market recklessness of his predecessor. Just as Bush envisioned a "kinder, gentler nation" to signal his commitment—at least in rhetoric—to a broader social ethic, he appropriated environmental crisis imagery to move beyond Reagan's open hostility to environmental values. Yet, just as Bush's calls for "a thousand points of light" emphasized voluntary action over social policy, popular environmentalism focused on personal responsibility and green consumerism as solutions to the crisis.

Indeed, as Earth Day 1990 approached, neoliberal visions of environmental citizenship became increasingly dominant. While historians and political commentators often focus on the partisan divides between Democrats and Republicans, this emphasis on the differences between conservatives and liberals obscures the overall patterns of change—the retrenchment of the welfare state, the upward redistribution of wealth and income, the faith in free-market solutions—that have characterized American politics in recent decades. To understand the prospects and limits of popular environmentalism, we need to consider how neoliberalism shaped the movement—and its media representations—during this pivotal moment.

The term neoliberalism refers to the revival of free-market liberal ideas—the classic eighteenth-century liberalism of Adam Smith given new life in our own time. Emphasizing deregulated markets, privatization, and a diminished welfare state, the neoliberal ethos has been widely shared and promoted by both Democrats and Republicans from the late 1970s through today. During the neoliberal age, American political culture has become increasingly defined by skepticism toward government's ability to redress social problems and by rising faith in market solutions. Popular environmentalism intersected in complex ways with the neoliberal project, sometimes challenging the principles of deregulation but also prescribing individual, consumerist remedies to the environmental crisis.[6]

According to the logic of green consumerism, the market could be harnessed to save the planet. Individual economic actors, choosing to optimize their ecological virtue, could signal their environmental concern to corporations. These powerful, profit-seeking institutions would, in turn, respond to this growing demand by producing and marketing items that would allow consumers to feel that their purchases had a salutary impact on the environment. Green consumerism thus made shopping synonymous with politics. This vision narrowed environmental citizenship to be an expression of consumer choice rather than an effort to challenge corporate structures or renew public life. The strategy aimed to empower the consumer by framing the market as an essential realm of freedom, perhaps even the most effective site of democracy.

In the time leading up to and following Earth Day 1990, a wide array of media—including TV news, Hollywood film, and popular magazines—disseminated disturbing images of crisis and seemed to promote a new era of environmental responsibility. Rather than emphasizing the state's obligation to protect the citizenry from harm, though, the mass media, together with certain environmentalists, imagined individual consumers as the prime agents of change. Even as this new wave of environmental crisis coverage pushed back against free-market ideology, then, it did not represent a full-scale rejection of Reaganism. Instead, this period marked a transition to green neoliberalism, defined by the historian Ted Steinberg as "the idea that market forces combined with individuals all doing their part can save the planet."[7]

Media images played a crucial role in codifying this idea. In an age when market metaphors became increasingly dominant in American public life, faith in personal responsibility and market-based solutions also took center stage in popular environmentalism. The private sphere became the main site of environmental action, the place where virtuous consumers could atone for their ecological sins through recycling and other individual acts. At times, popular visual imagery exposed the reckless irresponsibility of certain corporations and helped Americans glimpse the environmental effects of consumer capitalism. Yet the crisis-oriented coverage ultimately moved from corporate malfeasance to visions of individual hope and consumer-induced change. In the neoliberal age, media packaging of environmental hope often framed the movement as a form of therapy, a way for individuals to cope with the distressing imagery of environmental crisis.[8]

Six weeks before the 1980 presidential election, in which Ronald Reagan would soundly defeat Jimmy Carter, *Time* magazine ran a cover story titled "The Poisoning of America." The ghoulish cover image showed a white person immersed in contaminated water, on the verge of drowning as lethal liquid rose to the tip of the nose (fig. 12.1). Beads of sweat gathered on the victim's face, signifying broader public anxiety about "those toxic chemical wastes." While the brilliant sky and verdant trees connoted ecological harmony, below the water's surface the human and piscine victims had become toxic skeletons. "We wanted to show that aboveground, things may look O.K., but underneath it's death," the artist James Marsh explained. Even as Reagan and other conservative politicians pushed for deregulation, the fear of subterranean toxicity persisted throughout the 1980s.[9]

Time's cover image emerged out of the broader public anxiety associated with the Love Canal crisis. In 1978, Lois Gibbs and other neighborhood residents in Niagara Falls, New York, learned that the former Love Canal site—upon which their homes and children's school had been built—contained twenty-two thousand tons of chemical wastes, stored in metal drums that were beginning to leak. For the next two years, TV news and other media sources devoted considerable attention to the crisis and used familiar tropes to express a sense of danger, especially to children. As the sociologist Andrew Szasz observes, this TV coverage "conveyed the ominousness of the events at Love Canal by showing visuals that seemed to signify 'normalcy,' but undermining or reversing, signifying the opposite, through voice-over narration." In one example, he notes, "a boy bicycles along a quiet suburban street while the narrator says, 'There have been instances of birth defects and miscarriages among families.'" In another, "kids play on a community

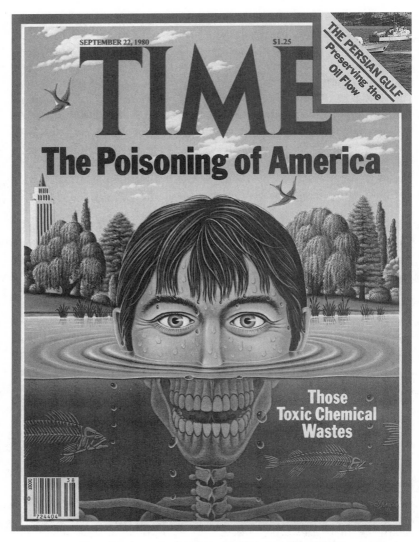

FIGURE 12.1. *Time* magazine cover, September 22, 1980. Copyright 1980 Time Inc. Used under license. *Time* and Time Inc. are not affiliated with, and do not endorse products or services of, licensee.

playground while the narrator reports the New York Health Department's recommendation that pregnant women and children evacuate." Much like the SANE ad that featured a picture of smiling children but undercut their giddiness with the textual warning, "*Your* children's teeth contain Strontium-90," these TV reports used the narrator's words to disrupt the presumed normalcy of the visuals. Like previous examples of environmental crisis imagery, this focus on white children and pregnant white women conveyed universal victimhood and called upon the state to protect innocent citizens from harm.[10]

As leader of the Love Canal Homeowners' Association, Lois Gibbs strategically engaged with the visual media. Gibbs and other female activists emphasized their role as nurturing mothers to argue that the area's toxic leakage produced harmful levels of reproductive risk: from inordinate rates of miscarriages and still births to birth defects, illnesses, and possible genetic damage among area children. Love Canal activists gained media sympathy by mobilizing images of mothers and children to convey concern for futurity and demand that the state protect young citizens and potential life from toxicity. The media, Gibbs later explained, "'loves women and children, especially the visual media.'" From newspaper and magazine photographs of children to heart-wrenching TV interviews with Love Canal parents, the media emphasized childhood vulnerability to heighten public fear of toxicity (fig. 12.2).[11]

Toxic waste entered the popular lexicon following Love Canal. Media reports suggested that potential Love Canals existed across the United States: many other communities might face the terror of buried chemicals oozing into the ground, contaminating the water and seeping into homes. *Time's* 1980 cover story confirmed the sense of universal danger by depicting the toxic threat as "The Poisoning of America."[12]

Even as the Reagan administration assailed the EPA and other regulatory agencies, free-market enthusiasts could not fully shake the legacy of Love Canal. For a while, the administration refused to comply with Superfund—federal legislation passed in the wake of Love Canal that provided for the cleanup of hazardous waste sites. Yet a major EPA scandal led to a political outcry against the administration's handling of the toxic waste issue. Even though Reagan successfully drove through his deregulatory agenda across a variety of government agencies, thereby gutting consumer, environmental, and workplace protection, toxic waste proved to be an important exception. "With toxic waste," Szasz explains, "'deregulation' ceased to be an abstraction to be vaguely ap-

FIGURE 12.2. Lois Gibbs and her children, Missy and Michael, carry signs at a Love Canal protest at the Niagara Falls, New York, city hall, 1978. Photograph by Penelope Ploughman. Copyright 1978, all rights reserved. Courtesy of University Archives, State University of New York at Buffalo.

proved as contributing to economic improvement. Deregulation was, instead, experienced as the very real dread that one could find oneself in a position like that of those people at Love Canal."[13]

For years to come, Love Canal remained an environmental icon and played a crucial role in advancing environmentalist calls to strengthen hazardous waste legislation. Despite his personal opposition to these measures, Reagan felt compelled, first in 1984 and again in 1986, to sign into law toxic waste bills that expanded this dimension of the environmental regulatory state. The continued saliency of Love Canal and popular fear of hazardous waste forced this concession. Through-out the 1980s, the media repeatedly warned of "The Poisoning of America." *Time*, for example, in "an unprecedented step for the maga-zine," ran a cover image in 1985 almost identical to the one it had run in 1980, except that now the toxic waters reached even higher, all the way up to the person's ears.[14]

Still, even as the federal government promised to dispose of hazard-ous wastes safely and to clean up Superfund sites, many environmental-ists charged that the entire regulatory system was fundamentally flawed in its effort to deal with toxicity. The major problem, from which other problems emanated, was that federal policy did not mandate "waste minimization" for corporations, because the US government refused to "interfere with the production processes of enterprises." By not requir-ing massive reductions in toxicity, levels of hazardous waste production did not decline over time. Instead, the escalating amounts of chemical contaminants were handled via "end of the pipe" remedies, such as the construction of hazardous waste landfills and incinerators, especially in poor and minority communities. These policies exacerbated ecological injustice within the United States by burdening marginalized groups with greater exposure to toxicity. This regulatory system also under-wrote the rise of the hazardous waste trade, in which wealthy countries such as the United States began to export toxic wastes to developing countries. By shifting waste disposal to distant environments, global dumping displaced environmental risk abroad and allowed waste gen-eration to continue apace. The focus on Love Canal and the media's continual emphasis upon the white ecological body under assault again indicate the limits of media representation: as with the gas mask and other emblems of environmental risk, the unequal distribution of pol-lution and toxicity remained hidden from view.[15]

While the media gave public visibility to the toxic waste issue and suggested that the state needed to protect the citizenry from harm, this coverage operated in tension with other depictions of environmentalism that imagined the political world in an individualist frame and thereby reinforced neoliberal models of citizenship. By promoting the notion of governing at a distance, by instilling a sense of guilt and personal re-sponsibility for the environment, the media framed recycling and other individual actions as a form of therapeutic relief, a way for consumers to feel better—both about themselves and the environmental future.

At the beginning of the award-winning, critically adored independent film *Sex, Lies, and Videotape* (1989), Ann (played by Andie McDowell) talks to her therapist about garbage. "Garbage," she announces. "All I've been thinking about all week is garbage. I mean, I just can't stop thinking about it."[16]

"What kind of thoughts about garbage?" he asks.

"I've gotten real concerned over what's going to happen with all the garbage," she explains. "I mean, we've got so much of it. . . . We have to run out of places to put this stuff eventually. The last time I started feeling this way was when that barge was stranded, . . . and nobody would claim it. Do you remember that?"

Ann was referring to the *Mobro*, the garbage barge that set sail from Long Island, New York, in March 1987, looking for a place to dump its cargo of more than three thousand tons of trash. The *Mobro* journeyed down the Atlantic Coast and then over to the Gulf of Mexico, only to be rejected by six different states and three foreign countries before returning to New York two months later (fig. 12.3). Television news broadcasts provided viewers with regular updates on the "vagabond barge," which became, according to CBS, "the most watched load of garbage in the memory of man." Johnny Carson lampooned the *Mobro* on *The Tonight Show*, while Phil Donahue hosted part of his daytime show aboard the garbage-laden vessel. Meanwhile, a *New Yorker* cartoon captured the absurdity of the *Mobro*'s voyage: Two rats stand in front of its towering bales of trash. "In your wildest dreams," one asks the other, "did you ever think you'd one day be cruising the Gulf of Mexico?" (fig. 12.4).[17]

The *Mobro* became a news icon that generated popular concern about the garbage crisis. As they followed the wandering barge to the Gulf of Mexico and back, TV news, popular magazines, and other media sources also began to devote increasing attention to trash. This coverage framed the *Mobro* as a harbinger of things to come: a sign of the looming landfill crisis and the need to promote recycling efforts across the United States. As the *New York Times* editorialized, "all states and localities ought to recognize the homeless scow as a floating Paul Revere. It has sounded an alarm about an imminent threat to American life posed by the vast tonnage of waste the nation produces."[18]

The *Mobro* helped visualize the ecological consequences of consumer culture. While corporate advertising and the mass media typically obscure the environmental effects of consumption, this imagery revealed the mounting levels of trash in contemporary America. The garbage crisis acquired a sense of urgency and seemed to demand immediate action. Yet, rather than envisioning alternatives to the systemic production of so much waste, mainstream framings of this

FIGURE 12.3. *Mobro* garbage barge, 1987. Photograph copyright Rick Maiman / Sygma / Corbis.

"In your wildest dreams, did you ever think you'd one day be cruising the Gulf of Mexico?"

FIGURE 12.4. *New Yorker* magazine cartoon by James Stevenson, 1987. Used by permission of the Cartoon Bank / New Yorker Collection.

crisis merely embraced recycling as the solution to the garbage glut. Long promoted by environmentalists, but rarely implemented by local and state governments, recycling programs finally became more common during the years surrounding Earth Day 1990. Indeed, as cities and towns across the United States adopted recycling programs, and as blue boxes by the curb eventually became ubiquitous, recycling moved from the margins of American life to the mainstream. Recycling became the paradigmatic expression of environmental concern, an everyday action that promised to save the planet. By concentrating public attention on individual action, not on wasteful corporate practices or inadequate regulation of industry, recycling also indicated the limits of popular environmentalism. As part of a broader process of environmental privatization, recycling made the home into the crucial site of environmental change.[19]

This focus on the domestic realm as the locus of environmentalism often presented women as being especially responsible for adopting green practices and ensuring that their homes demonstrated ecological virtue. Despite the important gains of feminism, advertisers still targeted women as the nation's leading shoppers and therefore the main participants in green consumerism. In contrast to Lois Gibbs's politicized move from the private to the public sphere, these marketing appeals confined women (and environmentalism) to the domestic realm. Green consumerism encouraged individual women (and men) to internalize guilt and a sense of personal responsibility for the deepening environmental crisis. Recycling and other individual remedies all promised to provide therapeutic relief through personal action.[20]

In May 1990, just after the twentieth anniversary of Earth Day, the popular women's magazine *Glamour* claimed that millions of American women were feeling much like Ann in *Sex, Lies, and Videotape*, worrying about garbage and other troubling signs of environmental crisis. *Glamour* asked: "Hey, Ann. Are you feeling overwhelmed by garbage glut? By acid rain and global warming? By shrinking rain forests and toxic waste? Guess what: You're not alone. Women all over the country these days are thinking, fretting, talking, arguing, dreaming, and, yes, obsessing about the planet's health." The article emphasized the personal, gendered dimensions of ecological responsibility, suggesting that women possessed unique environmental agency and were perfectly suited to act as the planet's caretakers. "They're feeling guilty about bad habits and bragging about their good ones," *Glamour* continued. ". . . And discovering how it feels to know you're deciding the planet's fate, day by day." *Glamour*, like other mainstream sources, personalized the environmental cause to emphasize micro-solutions that women, in particular, could implement in the home. Yet this widespread focus on individual action in the private sphere would also be joined by increasing media attention to the global dimensions of the environmental crisis, to the macro-view of planetary peril that repeatedly circulated in the years leading up to Earth Day 1990.[21]

This macro-view presented a cluster of environmental problems—ozone depletion, global warming, and rain forest destruction—as transnational and interconnected. By the late 1980s, these separate problems merged into an all-encompassing vision of a global environmental crisis.[22]

Media coverage of ozone depletion proved crucial in establishing the notion of planetary peril and in challenging Reaganite dreams of boundlessness. Public concern about this issue began in the 1970s, when reports of aerosol spray cans releasing chlorofluorocarbons (CFCs) into the atmosphere led to dramatic changes in consumer behavior and government policies. The ozone-destroying properties of CFCs ultimately threatened humans with increasing skin cancer risk, and thus generated popular fear about this issue. While CFCs were largely banned from spray cans, their use in other products—including refrigeration, air conditioning, Styrofoam, and computers—continued to rise. In the mid-1980s, the media once again focused public attention on ozone depletion. Widespread reports of an "ozone hole" over Antarctica conveyed a sense of urgency, and made the issue appear as a crisis that demanded immediate action. NASA satellite photographs provided spectators with an eerie glimpse at the damage to the ozone layer. As one NASA official explained, "As soon as we made images, we could see it wasn't a small, isolated hole." Likewise, as these pictures circulated in the visual media, the American public learned how to see a global environmental crisis in the making. The satellite view, together with computer-generated graphics indicating the extent of ozone depletion, lent scientific certainty to the images and revealed the steadily worsening condition of the atmosphere.[23]

These media images pushed back against free-market ideology. By showing that industrial society destroyed the ozone layer and thereby imperiled human health, the imagery suggested that harmful substances needed to be controlled so that the environment could be protected from further damage. Reagan at first tried to delay government action on CFCs. However, a media controversy soon erupted when Secretary of the Interior Donald Hodel reportedly urged Americans to rely solely on personal protection measures—"staying indoors or relying on hats, sunglasses, and suntan lotions"—to ward off the menace of skin cancer. His position was ridiculed by environmentalists and the media, and held up as evidence of the administration's reckless disregard for the environment. Following this media debacle, Reagan suddenly began to push for stronger controls of CFCs and enthusiastically endorsed the 1987 Montreal Protocol, the pivotal international agreement that mandated the phase-out of CFCs and other ozone-depleting chemicals.[24]

The ozone hole was soon joined by other transnational environmental problems—especially rain forest destruction and global warming—to produce an overarching frame of global environmental crisis. The summer of 1988 marked a crucial moment in the history of environmental crisis imagery. That summer, hypodermic needles, vials of blood, and other medical waste washed up on the Atlantic Coast. The ghastly spectacle of medical debris prompted numerous beach closings and, as one TV news broadcaster put it, "turn[ed] holidays into horror stories." Meanwhile, a catastrophic drought and heat wave desiccated large swaths of the Great Plains and Midwest. Scorching the soil from Canada to Texas, ruining crops and destroying farmland, the great drought of 1988 became the costliest natural disaster in US history up until that time. Record-setting temperatures across the United States, including "an unprecedented number of 100° days" in many places, also became major news stories that summer. The sweltering heat wave, together with stunning images of fires raging in the Amazon rain forest as well as in Yellowstone National Park, helped advance the idea of an interconnected crisis and contributed to the popular perception of global ecological collapse.[25]

The drought and heat wave suddenly made the scientific theory of global warming newsworthy. Although many scientists had been warning about the greenhouse effect for years, the issue attracted more media attention during the summer of 1988 than ever before. *Time*'s cover story on the drought, titled "The Big Dry" and published on Independence Day, asked readers to consider whether the "latest dry spell" also represented "an ominous harbinger of things to come." Likewise, a week later, *Newsweek*'s cover story, titled "The Greenhouse Effect," linked the disastrous drought to scientific forecasts of future warming. Media temporality—the focus on the immediate conditions of the moment—brought public visibility to climate change and encouraged audiences to view the weather in long-term perspective—to consider the summer of 1988, with its "images of blasted crops and burning forests," as "a warning signal, a visible preview for what the future might hold."[26]

By the time George Bush boarded a ship in Boston Harbor, the environmental crisis had thus become one of the leading news stories of the year, one that linked the local to the global and which depicted a multiplicity of issues as interrelated, as converging images of worldwide ecological danger. The media pictured an array of problems as together constituting a vision of planetary peril.

In January 1989, *Time* magazine broke from its tradition of naming an individual the "Man of the Year" and instead placed the "Endangered Earth" on the cover. "This year the earth spoke, like God warning Noah of the deluge," one editor explained. "No single individual, no event, no movement captured imaginations or dominated headlines more than the clump of rock and soil and water and air that is our common home."[27]

Time asked the artist Christo to produce a cover image that would represent "earth's vulnerability to man's reckless ways." Renowned (and sometimes reviled) for his large-scale earthworks and public art projects, Christo decided to take a small globe, wrap it in plastic, bind it in rope, and then photograph the object on a beach at sunrise (fig. 12.5). Titled *Wrapped Globe*, the piece drew upon other popular earth imagery, including the iconic whole earth photograph. Unlike these documentary images, though, *Wrapped Globe* called attention to its artifice. Rather than a shining blue ball that inspires awe in spectators, the plastic-wrapped, bundled-up globe appears as a synthetically produced item confined by human society. The piece, according to one commentator, "vividly warns us that the earth is not a package that can be exchanged for a new globe at the local outlet store."[28]

Time's special issue featured dramatic images—rain forests burning in Brazil, the ozone hole over Antarctica, cracked earth in the Great Plains, the *Mobro* returning to New York—that together signified an overarching sense of environmental crisis. The magazine presented the images as interconnected, as portents of planetary disaster. The articles likewise gestured toward a global vision of environmentalism, a campaign to unite people around the world in a common project to rescue the Earth. "Now, more than ever," *Time* explained, "the world needs . . . a universal crusade to save the planet." Later that year, with the decline of the Cold War and the dismantling of the Berlin Wall, global environmental consciousness became even more prominent in popular framings of environmental issues.[29]

FIGURE 12.5. *Time* magazine cover, January 2, 1989. Copyright 1989 Time Inc. Used under license. *Time* and Time Inc. are not affiliated with, and do not endorse products or services of, licensee.

While many letter writers praised *Time* for raising public awareness, conservative ideologues rebuked the magazine and other media outlets for environmental coverage that, they charged, violated the rules of journalistic objectivity. Led by *Wall Street Journal* columnist David Brooks, they rejected *Time*'s eco-partisanship and dismissed the entire discourse of environmental crisis as "a smorgasbord of apocalypse." "Somehow the idea has gotten around that the environment isn't a normal political issue, but a quasi-religious crusade," Brooks claimed. "As a result, public discussion of the environment has been about as rigorous as one expects from a jihad." According to Brooks, the *Time* cover story represented just one example of a much broader media embrace of the environmental cause. Newspapers, magazines, and TV networks, he argued, all conspired to scare the public with apocalyptic visions. Powerful media interests became environmental advocates by forsaking objectivity and peddling alarmism instead.[30]

Yet Brooks need not have worried so much about green going mainstream. Even as popular magazines, TV news, Hollywood films, and other media sources presented environmentalism as a cause that everyone should embrace, this coverage repeatedly focused on the personal dimensions of change. Popular environmentalism thus appeared as a therapeutic, consumer-based response to ecological danger. Just as media coverage of 1970s energy crises had projected fear of nuclear accidents but failed to grapple with soft-path solutions, the visual media during this period emphasized dramatic spectacles of danger and defilement, but did not engage with broader structural ideas about how to reduce toxicity, diminish greenhouse gas emissions, and envision alternatives to industrial agriculture. Just as dominant media frames in the 1970s had emphasized personal conservation measures rather than collective visions of an alternative energy future, this newer coverage similarly—but much more relentlessly and confidently—envisioned consumers as being endowed with tremendous power, as individuals who could use their purchasing decisions to save the planet. More than ever before, green consumerism and other forms of personal action became the crucial levers of change, the site where hope resided, the realm in which individuals could simultaneously express their environmental values and gain a sense of therapeutic relief.

The following three chapters will consider three image events—the environmentalist campaign to ban Alar, a carcinogen sprayed on apples; the 1989 *Exxon Valdez* oil spill; and the media packaging of Earth Day 1990—as pivotal examples of environmental spectacle in a neoliberal age. In both the Alar scare and the *Exxon Valdez* spill, media temporality masked the systemic implications of environmental crisis to emphasize the spectacle of dramatic catastrophe and immediate risk. Meanwhile, Earth Day 1990 demonstrated how environmental feelings had become channeled into a narrow, privatized vision that confused therapy for protest, that made caring for the environment synonymous with political engagement, and that branded ecological concern as a gratifying form of virtuous consumption.

In some cases, the producers of popular environmental images consciously recognized and promoted the therapeutic value of individual action. Consider the Ad Council's recycling campaign, produced in conjunction with the Environmental Defense Fund (EDF), a mainstream national group. Through a series of TV spots and print ads that circulated from 1988 to 1992, the campaign merged the macro-view of crisis with the micro-solution of recycling to encourage personal action to save the planet. The Ad Council used the iconic photograph of the whole earth to visualize the global meanings of individual action (fig. 12.6). "If you're not recycling," the text explained, "you're throwing it all away." Much like the Crying Indian, the "Don't Be Fuelish" campaign, and other previous Ad Council representations of environmentalism, the recycling campaign again emphasized personal responsibility. More than ever before, though, this imagery positioned the individual consumer within a global framework of crisis to suggest that the viewer's failure to recycle constituted a spectacular threat to the planetary future.

The Ad Council and EDF began developing this campaign soon after the *Mobro* story. "It used to be that we rarely gave a thought to our trash after we placed it at the curb to be picked up and taken away," the Ad Council explained. "Now, like the sequels to a popular horror movie, our discarded waste keeps returning to haunt us."[31]

To assuage this anxiety, the campaign aimed to present recycling as an everyday action that would offer consumers green therapy and a feeling of personal involvement in sustainability. In promoting the campaign, EDF President Frederic Krupp emphasized the growing sense of emotional paralysis, the feeling of being overwhelmed by depressing imagery of environmental crisis. "Acid rain . . . global warming . . . toxic waste . . . rainforest destruction," he began. "There are so many urgent, complex environmental issues that need our attention. While polls show Americans care deeply about the environment, a common sentiment is: 'Sure I'd like to help. But what can one person do?'" According to the Ad Council's marketing strategy, the campaign would target spectators who felt "daunted by the apparent size and complexity of environmental problems." The ads would "combat this overwhelming feeling by offering people the knowledge that they can individually have a positive impact on the environment." Audiences would learn that recycling "is emotionally rewarding" because it was

A little reminder from the Environmental Defense Fund that if you're not recycling, you're throwing away a lot more than just your trash.

You and your community can recycle. Please write the

© 1988 EDF

Environmental Defense Fund at: EDF-Recycling, 257 Park Avenue South, New York, NY 10010, for a free brochure that will tell you virtually everything you need to know about recycling.

ENVIRONMENTAL DEFENSE FUND CAMPAIGN
MAGAZINE AD NO. EDF-2800-90—7" x 10"—4/C (120 Screen)
Volunteer Agency: Deutsch, Inc., Campaign Director: Harry E. Davis, DuPont Co.

FIGURE 12.6. "If you're not recycling. . . ." Advertising Council / Environmental Defense Fund advertisement, 1988–92. Courtesy of the Ad Council Archives, University of Illinois, record series 13/2/215.

"the most direct way [they could] contribute to the protection of the environment."[32]

Rather than dismissing the Ad Council campaign and other popular images as merely a distraction from systemic environmental problems, we need to consider how they performed important cultural and political work during this period: how they constructed templates for neoliberal citizenship by addressing consumer anxieties about the environmental future. Once again, the Ad Council placed lopsided faith in the efficacy of individual action. However, unlike the Crying Indian and "Don't Be Fuelish" campaigns, which had angered and alienated many environmentalists, the recycling campaign was sponsored by a leading environmental organization.[33]

Why did this vision of consumerist environmentalism become so widely represented and enshrined during this particular historical moment? Three reasons stand out. First, the end of the Cold War, together with the success of the Alar campaign and other green consumerist projects, all reinforced the increasing popular faith in the capitalist market as an instrument of democracy. The events of the period surrounding Earth Day 1990 would not only lead to the collapse of the Soviet Union but also reveal haunting visual evidence of extensive environmental damage throughout the Eastern Bloc. To many American viewers, this imagery offered incontrovertible proof of the inherent superiority of US capitalist democracy and contributed to popular interest in what Francis Fukuyama exuberantly called "the end of history." Meanwhile, the fall of the Berlin Wall and the iconic image of the lone Chinese protester facing down a column of tanks near Tiananmen Square galvanized popular hope in the capacity of people to change the world. Like the Tiananmen Square photograph, the media spectacle of Earth Day 1990 and green consumerist strategies all zeroed in on the individual as the agent of change.

Second, the visual politics of Earth Day 1990 oscillated between the global and the personal to create a neoliberal politics of scale. At the macro level, the environmental crisis seemed both worldwide and potentially apocalyptic. Just as *Time* placed the "Endangered Earth" on its cover, other media imagery expressed a global vision of environmental crisis, a threatening condition that put all people and ecosystems at risk. While these texts produced a sense of universal vulnerability, they also obscured power relations and environmental inequities, both within the United States and globally. Even as minority communities struggled against environmental racism in America, and even as the United States increased its export of toxic waste to the developing world, these images presented all humanity as linked by a common, equally shared geography of risk. This global sense of crisis would be joined by a relentless focus on the micro, on the individual as the agent of change, the enlightened consumer who can redirect markets to ensure sustainability. From ecological systems in peril to choice-making individuals saving the planet, media framings of Earth Day applied micro-solutions to a macro-crisis. Neoliberal models of citizenship mediated this apparent contradiction between planetary crisis and personalized solution. The neoliberal emphasis upon individual responsibility and choice-making consumers imagined environmental citizens governing at a distance, demonstrating their ecological concern in the private realm.

Third, the rise of recycling programs provided a tangible way for individuals to gain a sense of involvement in the environmental cause, to feel that their personal actions contributed to ecological improvement. It is not coincidental that recycling—above all other actions— became the most prescribed form of environmental citizenship during the period surrounding Earth Day 1990. While modern recycling had emerged as a countercultural practice in the late 1960s and early 1970s, the mainstreaming of recycling did not happen until twenty years later as beverage and packaging companies, together with local, state, and federal government agencies, began to tout it as the panacea to the nation's growing solid waste crisis. The recycling logo, developed in the aftermath of Earth Day 1970, suddenly became ubiquitous, appearing on bins and beverage containers as an emblem of sustainability, a reminder for consumers to take personal responsibility for the planetary future.

The period surrounding Earth Day 1990 thus constitutes not only a crucial environmental moment, but also a pivotal phase in the emotional history of capitalism. Capitalism and emotional politics became enmeshed in new and complicated ways during this period: while corporations sometimes seemed responsible for the environmental crisis, the market, spurred on by enlightened consumers, also seemed to offer the most promising path out of the abyss. Mainstream portrayals of environmentalism addressed the emotional needs of certain consumers— especially a relatively privileged, primarily white demographic—but ignored the power dynamics and racial dimensions of environmental risk. Like other neoliberal practices, green consumerism offered a fantasy of empowerment for affluent Americans. By equating economic choice with political power, this vision of popular environmentalism reinforced the inequities of neoliberal citizenship and promulgated hope in short-term consumerist solutions to environmental problems.[34]

THIRTEEN

Meryl Streep, the Alar Crisis, and the Rise of Green Consumerism

In 1989, Meryl Streep became an environmental celebrity. It was an unusual role for the "intensely private actress," who had always sought to keep her family out of the limelight and had refused to become, in her words, "celebrity junk food." Streep took on this role after learning that the Natural Resources Defense Council (NRDC) had completed a report documenting the presence of carcinogenic chemicals in the nation's food supply. "When I read this report," Streep explained, "the evidence was so stunning and compelling that all my reticence about getting involved went away. . . . I'm not just concerned; I'm furious! I'm in the position to give my children anything they want, but I can't provide them with wholesome food. Something's fundamentally wrong." Streep agreed to become the spokesperson for an NRDC campaign to publicize the risks of pesticides and chemicals applied to food, especially the danger of Alar-laced apples consumed by young children. Like many environmental activists before her, Streep sought to merge the private and public realms by bringing parental feelings into politics.[1]

For the NRDC, Streep's participation promised to heighten the visibility of the campaign, to confer additional newsworthiness on Alar through the presence of the revered actress. From Dr. Spock warning about radioactive fallout to Jane Fonda protesting nuclear power and Robert

Redford promoting solar energy, celebrities had long been involved in popular environmentalism and had helped bring media attention to various campaigns. During the late 1980s and early 1990s, though, the phenomenon of celebrity environmentalism—especially among Hollywood stars—became a media story in its own right. As *People Weekly* noted in 1990, "every peril on the planet seems to have acquired a star to call its own." Streep's role in the NRDC campaign arguably marked the most successful example of a movie star influencing environmental politics: Three months after she entered the environmental spotlight, Alar would be removed from the US market, thus protecting consumers from its menacing toxicity.[2]

The interplay between Streep's public image and private life, between her star power and status as a concerned mother made her the ideal spokesperson for the NRDC campaign. Throughout the 1980s, media coverage of Streep emphasized her effort to combine career with motherhood: she was "Magic Meryl," "the most-sought-after actress in Hollywood," but also "a working mother on a grand scale" who "simply will not sacrifice her family life for career." By becoming a spokesperson for the NRDC, Streep demonstrated concern for her children's health by entering the public realm as an environmental activist.[3]

Streep began playing this part in conjunction with the NRDC's release of *Intolerable Risk: Pesticides in Our Children's Food*. Although the report examined over twenty different chemicals used in agriculture, one specific claim was extracted by the NRDC and emphasized by the news media: *Intolerable Risk* identified Alar, the trade name of the chemical Daminozide, as "the greatest source of cancer risk" and warned that "apple products, including fresh apples, apple sauce and apple juice" exposed children to dangerous quantities of this carcinogen.[4]

In the NRDC campaign, the emphasis on the vulnerable bodies of children attracted considerable media attention but reduced the issue to a single chemical applied on a single fruit rather than to systemic questions about the environmental hazards of industrial agriculture. Streep's performance as a real-life mother and political activist shows how the mass media offered an arena in which environmentalists could contest the conservative effort to deregulate industry by mobilizing popular fears about toxic chemicals. Yet this campaign also revealed the difficulty of presenting and circulating alternative visions of the future, scenarios that moved beyond isolated instances of reform to imagine large-scale reductions of toxicity. The Alar campaign thus indicated the limits of the environmentalist engagement with the mass media. Too often, the movement itself became a kind of consumerist spectacle, using fear to focus public attention on specific hazards, but failing to communicate more far-reaching ways to confront the proliferating presence of toxins in the environment and in people's bodies. The Alar crisis generated a sense of urgency, made this particular chemical seem uniquely threatening, but ignored systemic critiques of industrial agriculture as well as radical proposals to reduce the escalating production of toxicity.

The Alar campaign also marked a pivotal moment in the history of green consumerism. Streep and the NRDC positioned consumers, especially the young, as the prime victims of pesticide exposure. The campaign sought to mobilize parental concern to demand that the US government protect its youngest citizens from the perils of Alar and other toxic agents in food. Alar's demise, though, came not from government action but rather from the decision of its manufacturer—responding to public concern and declining apple sales—to banish it from the US market. If American consumers were victims, they also seemed, in this case at least, to be neoliberal agents of change, using their purchasing decisions to alter corporate policy and create a healthier environment. This popular vision of environmentalism denied power relations and exaggerated the ability of individual economic actors to shape corporate priorities and patterns of resource use. Presenting the market as a realm of freedom where consumers could redirect production decisions by choosing to buy or not buy particular items, green consumerism seemingly became synonymous with political empowerment.

On the evening before the release of *Intolerable Risk*, CBS's respected news program *60 Minutes* aired a story titled "A is for Apple." Seated before a backdrop image of a shiny red apple superimposed with a skull and crossbones (fig. 13.1), the correspondent Ed Bradley began ominously: "The most potent cancer-causing agent in our food supply is a substance sprayed on apples to keep them on the trees longer and make them look better. . . . And who is most at risk? Children, who may someday develop cancer from being exposed to this product."[5]

The *60 Minutes* story had been arranged by Fenton Communications, a public relations firm hired by the NRDC. David Fenton, founder and CEO of the firm, explained that the campaign specifically targeted television and other visual media. "Usually," he observed, "public interest groups release similar reports by holding a news conference, and the result is a few print stories. Television coverage is rarely sought or achieved." Fenton wanted to heighten the visual dimensions of the Alar story to generate an "intensity of exposure" that would be

FIGURE 13.1. "'A' Is for Apple." Frame capture from CBS television news program *60 Minutes,* 1989.

"uncommon in the non-profit world." "The symbolic appeal of the thought of apples as a threat to children," he believed, would provide the public with a compelling and easily comprehended image of environmental danger. As the *60 Minutes* background indicated, Alar drew upon cultural meanings associated with Snow White's poisoned apple. The shiny surface of Alar-laced products concealed their carcinogenic qualities; just as the wicked queen offered Snow White a deadly apple, Alar threatened children with its invisible toxicity.[6]

Fenton not only designed the campaign with these visual and cultural symbols in mind but also staged the release of the NRDC report to maximize media exposure. He granted *60 Minutes* the right to "'break' the story of the report in late February," just before it was given to other media outlets. He scheduled interviews "with major women's magazines like *Family Circle, Women's Day* and *Redbook.*" He also arranged for Streep, whose participation he described as an "essential element" in the campaign, to appear on *The Donahue Show* and other daytime programs. "Our goal was to create so many repetitions of NRDC's message that average American consumers . . . could not avoid hearing it—

from many different media outlets within a short period of time," he explained. "The idea was for the 'story' to achieve a life of its own, and continue for weeks and months to affect policy and consumer habits." Fenton's strategy, then, relied not just on the power of visual media to create evocative images but also on harnessing media temporality to generate a sense of urgency, to galvanize public concern through repeated exposure on TV, popular magazines, and other media forms. If the NRDC had not hired Fenton to orchestrate the campaign, the Alar crisis may never have happened.[7]

While the NRDC report primarily consisted of charts, tables, and scientific analysis, *60 Minutes* visualized and dramatized the Alar issue. As one commentator later remarked, "It seems noteworthy . . . that nowhere between the covers of [*Intolerable Risk*] is there a red delicious apple emblazoned with a skull and crossbones." The *60 Minutes* piece began with that image and then explained how the EPA's failure to deal with Alar in a timely fashion cast broader doubts upon its own effectiveness as the nation's designated protector of ecological bodies. Indeed, by the time the NRDC released *Intolerable Risk*, EPA studies had confirmed that Alar exceeded the government's tolerance limits for carcinogenic chemicals. Nevertheless, even as it acknowledged this risk, the agency claimed that laws forbade it from removing Alar from the market immediately. In one interview, Bradley challenged an EPA administrator for his agency's foot-dragging. Later in the episode, Janet Hathaway of the NRDC chastised the government: "If EPA doesn't think that the most potent cancer-causing chemical in our food supply is grounds enough to declare it an imminent hazard and remove it from food, well, I don't know what kind of risk it takes then to declare an imminent chemical hazard." The *60 Minutes* report helped the NRDC create a sense of urgency about banning Alar from the food supply.[8]

Yet the spectacle of risk generated by the broadcast—and by subsequent media coverage of the Alar crisis—did not accurately convey the long-term, accretive modes of danger emphasized by the *Intolerable Risk* report. In preparing the study, NRDC scientists used a "time-dependent model to estimate risk," a methodology that factored in the "long latency period" of toxins. The NRDC selected children's bodies as the focus of the study for both symbolic and scientific reasons. Fusing fact with feeling, cognition with emotion, the report deployed the trope of childhood vulnerability and examined the "lifetime cancer risk" posed by various chemicals. "Preschool children are receiving hazardous exposures to pesticides at the time when they are likely to be most susceptible to the toxic effects of these compounds," the report explained. The accumulating hazard of toxins in the human body would eventually result, the NRDC argued, in thousands of additional cancer cases during the lifetimes of Americans who were then the nation's youngest citizens. The NRDC projected an extensive time frame of risk to warn of the long-term effects of toxicity. If the government did not take action by banning Alar, then today's apple-juice-drinking children would—at some point in their lifetimes—face an increased cancer risk.[9]

The 60 Minutes report heightened the sense of danger by condensing time, framing Alar as an immediate risk to the vulnerable bodies of children. In one interview, Representative Gerald Sikorski, a Democrat from Minnesota, asked viewers to envision childhood cancer sufferers in relation to the Alar debate. "Go to a cancer ward in any children's hospital in this country," he said, "and see the bald, wasting-away kids and then make a decision whether the risks balance over the benefits." The camera then showed images of children drinking apple juice. Together, the interview and visuals posited a causal relationship between Alar and childhood cancer to deliver a frightening warning about the carcinogenic dangers lurking in their cups.

Soon after this broadcast, Streep appeared at a news conference and on Today, Donahue, and other popular daytime TV shows to announce the founding of Mothers and Others for Pesticide Limits as an offshoot of the NRDC. She also starred in two of the group's three TV commercials. In all these roles, Streep emphasized the role of mothers in protecting their children from harm. In one commercial she wears an apron and stands with a boy as they wash broccoli together (fig. 13.2). "The government says we should wash all our fruits and vegetables for two to three minutes," Streep states, followed by a meaningful pause. "With detergent. And then rinse them. Why? What's on the food? And why are children particularly at risk?" Another NRDC spot shows a succession of individual children eating fruits and vegetables in their high chairs. As the children appear on screen, the narrator lists various pesticides whose residues the children will presumably soon ingest and then asks, "Who invited them to dinner?"[10]

Streep, according to a quantitative analysis of media coverage, became "the most newsworthy figure in the Alar controversy," the individual most often quoted and discussed by the media. After the press conference and the release of the NRDC commercials, Streep testified in a Congressional hearing. Her comments, a CBS news broadcast claimed, "seemed to sum it all up." Streep asked the committee, "Are we

FIGURE 13.2. Meryl Streep and child. Frame capture from television advertisement by Mothers and Others for Pesticide Limits and National Resources Defense Council, 1989.

not allowed to know what's on the food? That's what's so distressing." She also visited the White House, hoping, as *People Weekly* reported, "to stop a bitter harvest" by "plead[ing] her case" with a Bush administration official.[11]

While the NRDC reached well beyond the issue of Alar on apples to consider a plethora of toxic chemicals in the food supply, the specific topic of Alar would be amplified in the popular media. The focus on Alar helped render visible the invisible threat of this particular chemical, but also obscured the other forms of slow violence directed against human bodies and ecological systems. NRDC director John H. Adams described the Alar campaign a "'one chemical at a time' strategy," an effort to publicize the danger of a specific substance. This emphasis narrowed discussion about the widespread presence of pesticides in agriculture to such questions as whether the government should ban Alar immediately and whether parents should allow their children to keep drinking apple juice.[12]

Indeed, throughout much of March 1989, TV news and other media dramatized the Alar issue, often reducing the larger pesticide debate to the following question posed by a CBS news report: "Are apples a health risk to children?" School districts across the nation—including those in New York City, Chicago, Los Angeles, as well as in hundreds of smaller cities and towns—answered affirmatively, as they decided to remove all apple juice and apple products from their school cafeterias.[13]

The apple soon became a contested symbol. Senator Joseph Lieberman, a Democrat from Connecticut, evoked biblical imagery as he decried the use of Alar: "If I may say so, not since the Garden of Eden has the apple caused such an uproar in our world." Just as Streep explicitly linked her status as a nurturing mother to the NRDC campaign, Lieberman deployed similar language to express his parental concern about Alar's potential effect on his young daughter. "Like many Americans, my wife and I . . . were shocked" by the *60 Minutes* story, he explained. "We went to our food cabinets and . . . threw out all the applesauce and apple juice we could find." In contrast, Steve Symms, a Republican from Idaho, demonstrated his complete confidence in the safety of American apples and urged everyone to "join me in eating an apple." In a dramatic gesture played back that night on TV news, he took a large bite out of a red apple while standing on the Senate floor.[14]

Consumer Reports magazine soon joined the debate by publishing a cover story that also evoked Snow White iconography (fig. 13.3). The magazine's intervention marked a departure from its tendency to act as "a buying guide to consumer durables." For its Alar story, *Consumer Reports* conducted an extensive study of apples and apple products and reported, with considerable dismay, that a large sample available in supermarkets contained high residues of Alar. In presenting its findings the magazine, like the NRDC, reaffirmed the right of consumers to increased protection from toxic threats. Republican Senator Slade Gorton denounced *Consumer Reports* for what he viewed as its deceptive use of apple symbolism that promoted the NRDC campaign: "A shiny red apple," he said, describing the magazine's cover image, "held by a macabre, black-fingernailed hand—the classic portrait of a wicked witch luring an innocent child to poison—all under the huge, red headline: 'Bad Apples.'"[15]

It would become a staple among the conservative media that the Alar crisis demonstrated the triumph of emotion over reason, hysteria over science. Soon after the NRDC released *Intolerable Risk*, the *Wall Street Journal* lambasted the study as another in a series of "witchcraft tales that have been drummed into the American psyche . . . by environmental groups." In much of this coverage, Streep would be derided as a hysterical woman who had no business addressing the public about scientific matters. The *Wall Street Journal* described Streep as "actress-

FIGURE 13.3. *Consumer Reports* magazine cover, May 1989. Copyright 1989 by Consumers Union of U.S. Inc., Yonkers, NY 10703-1057, a nonprofit organization. Reprinted with permission from the May 1989 issue of *Consumer Reports* for educational purposes only. No commercial use or reproduction permitted. www.ConsumerReports.org.

turned-toxicologist," clearly questioning her credibility and implicitly mocking her audience, composed largely of women, whom she addressed in "a string of talk-show appearances" in which she spread "public anxiety." Using the gender-inflected notion of hysteria, the *Wall Street Journal* and other conservative sources chastised Streep and claimed that media spectacles of crisis worked to benefit the NRDC.[16]

Media attention did contribute to the rejection of the chemical. As apples sales declined across the United States later that spring, Uniroyal, the manufacturer of Alar, decided to remove it from the market. One commentator explained, "Parents were not inclined to argue about how much Alar was tolerable; they wanted *none* of it in apples, and Uniroyal responded by an action to ensure exactly that." The emphasis on vulnerable bodies and threats to innocent children tarnished the reputation of Alar and buttressed the call for increased protection of citizen consumers. For the NRDC, then, the campaign resulted in a major victory: apples would no longer be laced with Alar, the chemical the group assailed as the most carcinogenic substance in the nation's food supply.[17]

In the aftermath of the Alar crisis, chemical industry leaders and right-wing commentators would repeatedly claim that the controversy demonstrated the failings of "our nation's environmentalism: an emotional reaction that's based on misinformation, disinformation and the faulty use of statistics." The entire episode, they argued, revealed the threat to reason and rational policy making posed by "movie stars brandishing pseudo-science and manipulating the media to dictate what we eat." Much like the conservative reaction to *The China Syndrome* and the Three Mile Island crisis, these industry spokespeople and right-wing pundits framed the Alar crisis as an example of environmental alarmism duping the media and the American public to fall prey to the movement's dubious claims and phony hysteria. Once again, they complained, image had trumped reality: media spectacle had colonized the domain of rational deliberation to produce a celebrity-dominated, overly emotional public sphere.[18]

Yet, even if the mass media contributed to the cessation of Alar's use, these same reports also obscured broader criticisms of the contemporary agricultural system. Rather than viewing the Alar episode in the dualistic terms of emotion versus reason, spectacle versus science, and image versus reality, we need instead to consider how the "one chemical at a time" strategy mobilized public concern over this specific substance but also worked to deflect attention from the long-term systemic problems associated with industrial agriculture, including the escalating use of pesticides. Indeed, in *Intolerable Risk* and other reports, the NRDC sought to promote long-term solutions to reduce pesticide use and proposed new policies to encourage a large-scale "transition from chemical-intensive agriculture to innovative, safer farming techniques." These measures included providing "growers with . . . financial protection" to "help ease the transition" to organic farming methods. NRDC reports also called for "a tax on pesticides" to "fund research and demonstration projects in alternative, low-chemical farming." While the NRDC thus envisioned major policy initiatives to foster a transformation in US agriculture, these proposals were completely ignored or dismissed in media coverage of Alar.[19]

To take a prime example, in an episode of the ABC news program *Nightline*, the host Ted Koppel seemed relatively sympathetic to the NRDC argument about removing Alar from the market quickly. Yet Koppel also framed the larger issue of pesticide use in a remarkably reductionist manner, warning that if farmers stopped using pesticides, then Americans could end up finding "worms in tomato sauce and insects in bread." In an interview with a NRDC leader, Koppel refers to the industrialization of agriculture over the past century and then asks her: "Would you pick the way things were 100 years ago or now?" When she refuses to take the bait and tries to offer a more complex response to his simplistic question, he interrupts her: "I want you to say what you don't want to admit—that things are better now." The episode foreclosed any meaningful discussion of alternatives to the current food system. *Nightline*, like other media broadcasts, elevated a moderate vision of reform but rejected more ambitious proposals developed by proponents of organic agriculture and sustainable farming. While Koppel agreed that Alar should be banned, the larger system of industrial agriculture, including its ever-increasing reliance on a battery of pesticides, went unchallenged. Koppel's interview conformed to a larger pattern, a media frame described by the agricultural writer Ward Sinclair as "the popular image of the organic farmer as a freakish health-food nut who is checking out of the twentieth century."[20]

Even though Alar coverage relentlessly focused upon the health dangers of one specific chemical and marginalized radical critiques of industrial agriculture, the crisis triggered increasing demand for organic produce, establishing a pattern that would be repeated in subsequent food scares. "Even today, the rapid growth of organic closely tracks consumers' rising worries about the conventional food supply," the writer Michael Pollan observed in 2001. "Every food scare is followed by a spike in organic sales." Often cited as a "watershed" in the history of organic food consumption, the Alar scare produced what *Newsweek* described as a "panic for organic."[21]

This consumerist response to the crisis marked an important example of what Andrew Szasz has termed the inverted quarantine approach to environmental problems. Spurred by environmental anxiety, increasing numbers of primarily affluent Americans seek to protect their bodies from toxicity by turning to a wide range of consumer products. Green purchasing decisions promise "individualized acts of self-protection," a way for consumers to isolate themselves from escalating environmental risks. The inverted quarantine approach thus indicates the class-based limits of green consumerism. By privileging individual safety over collective reform, neoliberal environmentalism provides protection only to those who can afford it.[22]

The Alar scare and the rise of green consumerism also worked to short-circuit time, to promise quick fixes to long-term problems. Consider the analysis offered by Dana Jackson, a co-founder of the Land Institute and a perceptive critic of industrial agriculture. While the episode "open[ed] an intense debate about . . . environmental risk," Jackson noted, it also obscured "the longer-term and systemic problems with agricultural chemicals in the environment" and ignored the broader "context of 'agribusiness,' the engine that drives so many American farms." The Alar scare thus followed a pattern set by media framings of *Silent Spring* and other pesticide debates. "Often such stories tend to make it seem as if, once some isolated problem has been fixed . . . , we can quit being concerned," Jackson continued. Focusing on specific dangers and short-term solutions, media imagery failed to grapple with the long-term destructive legacies of pesticide-dependent industrial agriculture.[23]

NRDC director John H. Adams took a markedly different lesson from the Alar campaign, which he described as "one of the epochal moments in NRDC's history" that "catapulted us to a level of public recognition and credibility we had never dreamed of." He emphasized that Alar's demise resulted not from an EPA action but from Uniroyal's. "The most lasting influence from the Alar case was the lesson it taught us about how we could best protect the public and influence the behavior of industry," he explained. "NRDC would no longer focus only on government, law, and regulation. From now on, we would also look to the marketplace and the powerful force of consumer choice." Rather than relying exclusively on government action, the NRDC would now put its faith in green consumerism.[24]

In the time leading up to Earth Day 1990, many other mainstream environmentalists would extol the beneficent power of the market, and would make individual consumer choice central to their vision of environmental sustainability. Besides the Alar case, they would point to other successful examples of consumer power, especially the boycott of canned tuna, launched in 1988 to protest fishing practices that killed dolphins.

In promoting the tuna boycott, the environmental group Earth Island Institute evoked the charismatic appeal of the dolphin, a creature often celebrated by nature films and marine mammal parks for its "affectionate image" and "human-like intelligence." Through a series of full-page newspaper and magazine advertisements, the group urged consumers to help stop "the dolphin massacre" by refusing to buy canned tuna. Other visual media dramatically increased public awareness of the plight of "trapped, drowned, [and] mangled dolphins." Both ABC and CBS News broadcast disturbing video footage, taken by Sam LaBudde, a scientist working undercover on a fishing vessel, of dolphins entangled in nets, "unable to breathe," "crying out," and eventually dying.[25]

Even the Hollywood film *Lethal Weapon 2* (1989) inserted a save-the-dolphin message. In one scene the actor Danny Glover, playing the part of a police lieutenant, is chastised by his family as he prepares to eat a tuna fish sandwich. "Tuna!" his children yell, registering their shock. "Daddy, you can't eat tuna," one daughter informs him. "I can't eat what?" he responds incredulously. "We're boycotting tuna, honey," his wife explains, "because they kill the dolphins that get caught in the net."[26]

A wide range of visual media—from surreptitiously obtained documentary footage to print media ads, TV news, and Hollywood film—all elicited sympathy for dolphins and helped persuade many viewers, including children and youth, to boycott tuna. Indeed, not just Danny Glover's fictional family, but actual schoolchildren across the United States badgered their parents, school boards, and leading tuna companies to stop the "dolphin slaughter." After seeing the LaBudde video, one student organized his classmates to write postcards to Star-Kist executives. Ten days before Earth Day 1990, at a press conference announcing that the company "would no longer buy, process or sell tuna caught at the expense of dolphins," its president read a number of these postcards, including one that asked, "How can you sleep at night knowing your company is doing this?" While Streep and the NRDC called for the protection of children's vulnerable bodies, the tuna boycott galvanized children and youth as activists who urged corporations to adopt more sustainable practices. Following the Star-Kist announcement, other leading tuna companies soon declared that they also would stop buying tuna caught in nets that ensnared dolphins. These companies, according to *New York Times*, were simply responding to "market forces," since they "surely couldn't allow [Star-Kist] to stand alone as the dolphins' only corporate champion." The market thus seemed to empower consumers, to give them a voice in corporate policies. As one commentator gushed at the time, "It's consumer power that will change the world."[27]

Both the Alar crisis and the tuna boycott would lend credence to the idea of green consumerism, and would provide hope that the capitalist market could mend its ecologically destructive ways. Yet these examples also revealed the limits to such a strategy for environmental politics. In each case, the visual media conveyed outrage and concern over a specific practice—the use of a particular chemical on one fruit, the type of nets used to catch one type of fish—to generate a sense of urgency and imply that if a particular practice were changed, then the crisis could be solved. For all their power to activate the emotions, for all their ability to document the environmental harm inflicted by a clearly identifiable agent, these visual symbols ignored the long-term ecological consequences produced by the structures of contemporary capitalism.

Even as they challenged corporate power, the Alar crisis and the tuna boycott reinforced the increasingly popular perception of the market as a frictionless, fast-moving realm that responded more quickly and effectively to citizen concerns than did the inefficient, red-taped domain of government. Indeed, the NRDC campaign against Alar was premised on the notion of EPA inertia, of bureaucratic delays that prevented the agency from banning this substance in a timely fashion. As we have seen, media coverage of environmental crises challenged the Reaganite faith in boundlessness and tempered the sense of limitless optimism that underwrote the antiregulatory agenda. Yet the legacies of these campaigns should not be understood within the narrow political frame of Republicans versus Democrats. The repeated emphasis upon green consumerist solutions to environmental problems instead demonstrates the increasing prominence of neoliberal citizenship practices. Propelled by media reports of Alar contamination and dolphin slaughter, consumers felt a profound sense of environmental anxiety and worried about the destructive impact of capitalist enterprise. These feelings, in turn, shaped their economic behavior, leading many Americans to stop buying apple products or to boycott tuna. The market's apparently rapid response—resulting in the end of Alar and the adoption of dolphin-safe nets—yielded important citizenship messages, suggesting that capitalism could redeem itself by attending to the environmental anxieties of consumers.

As Earth Day 1990 approached, green consumerism became central to popular expressions of environmentalism, and Hollywood would increasingly seem to align itself with the environmental movement. Much like the tuna boycott scene in *Lethal Weapon 2*, ecological messages would appear in numerous TV shows. Newspapers and popular magazines would profile the actions of an array of celebrity environmentalists: Sting, Madonna, and the Grateful Dead performing benefit concerts for the rainforests; Tom Cruise and others speaking at Earth Day rallies; and, of course, Streep warning about the Alar menace. "The '90s are the years for the environment," *USA Today* noted. "Ever wary of where the spotlight is shining at the moment, the glitterati are all but turning green as they realize the best place to be is marching in the environmental parade." Even as the mass media emphasized celebrity involvement in the movement, even as the presence of popular entertainers enhanced the newsworthiness of the environmental cause, this same coverage implied that the stars were merely "jumping on the bandwagon" and, as one critic put it, turning "political 'beliefs'" into "another consumer good." The crisis temporality of the media, together with the perceived trendiness of celebrities, concentrated public attention on environmental issues but often presented the movement as nothing more than a fashionable way of feeling and an enlightened form of consumption.[28]

Only a few weeks after Streep became an environmental celebrity, TV news, popular magazines, and other media would be filled with shocking pictures of the *Exxon Valdez* oil spill. As audiences glimpsed birds, otters, and other creatures suffering and dying, as they witnessed Alaskan beaches blackened by oil, many felt intense sadness for the threatened wildlife and visceral anger at the corporation and the reputedly drunk captain. Even more than the Alar crisis, the *Exxon Valdez* spill conformed to event-driven patterns of media temporality to make this particular episode seem uniquely catastrophic. As oil spill imagery unleashed feelings of sadness and anger, as more celebrities proclaimed their commitment to the cause, environmentalism became increasingly prominent in American public life. Yet many activists still wondered whether media cycles—and the visual emphasis on sudden environmental devastation—would ultimately work to advance or thwart their agenda.

FOURTEEN

The Sudden Violence of the *Exxon Valdez*

In late March 1989 Tony Dawson, a professional nature photographer, suddenly felt like a war photographer in-stead. On assignment for *Audubon* magazine, Dawson flew in a helicopter above the grounded *Exxon Valdez* super-tanker as it spewed millions of gallons of oil into Alaska's Prince William Sound. "The experience was unsettling," Dawson wrote, "but distantly so, like viewing television news reports of a foreign disaster. From above, the spread-ing crude and its coastal impact were impossible to judge." While the aerial view provided a broad, panoramic per-spective on the spill, Dawson wanted to use his camera to document in a closer, more intimate fashion the violence inflicted upon the region's environment. As he surveyed "the depth of the sound's wounds," Dawson felt "numbed by shock and sorrow." "During four days on the Barren Islands," he explained, "we found no live, oiled birds while counting nearly two thousand corpses deposited by the tides. . . . For two depressing days we heard the wet 'thunk' of steel against meat and bone as . . . crews used gaff hooks to gather bodies. . . . The spill created a grow-ing war zone, every day a new beachhead, every night a new body count."[1]

Dawson's photograph of a dead sea otter became one of the most widely seen images of the disaster (fig. 14.1). He found the animal washed ashore, "posed grotesquely" with its "paws curled against its chest [and] teeth bared." Dawson's close attention to the victim's bodily features

FIGURE 14.1. Dead sea otter following *Exxon Valdez* spill. Photograph by Tony Dawson, 1989. Used by permission of the Estate of Anthony Dawson.

and facial expression works to heighten audience pity. The image reveals a dramatic tension between life and death, an animated being turned into a lifeless specimen. The otter's paws, permanently immobilized by oil, serve as a poignant reminder of the animal's past mobility. Meanwhile, its mouth, with teeth gleaming against the petroleum-laden darkness, forms what could be viewed in anthropomorphic terms as a desperate scream by a creature doomed to die. The otter calls upon viewers to intervene and help put an end to the rising body count.[2]

Like the Santa Barbara oil spill twenty years earlier, media coverage of the *Exxon Valdez* accident presented a grisly spectacle of nature defiled. The sudden violence of the spill deprived animals of crucial bodily functions—made it impossible, for example, for otters to keep themselves warm. Audiences saw vivid, distressing evidence of what happened to otters as their fur became saturated with oil: the pitiful creatures began to shiver, and many soon died.

The disaster already seemed senseless, but on March 30, six days after the spill began, all three TV news programs began by describing its cause as downright obscene. On CBS, Dan Rather announced: "Good evening. Captain Joseph Hazelwood was legally drunk when his Exxon corporation tanker ran aground off the Alaskan coast." Over on NBC Tom Brokaw concurred, while ABC's Peter Jennings equivocated somewhat, noting that since Hazelwood had not been tested until more than ten hours after the crash, "it was difficult to know whether he was drinking before the accident, afterwards, or both." On previous nights, news coverage had decried the captain's drunken ways and the dubious decisions he made on the night of the crash. All three networks reported, with some astonishment, that Hazelwood had twice been convicted of drunk driving near his home on Long Island, New York, and had had his driver's license revoked three times. At the time of the accident, they all noted, Hazelwood could not legally drive a car but still could command a supertanker. When the *Exxon Valdez* struck Bligh Reef, though, the captain was not behind the wheel: he had gone below deck and left at the helm an unlicensed third mate, someone who was not even authorized to guide the ship through those waters. Media stories of the drunken captain and the unqualified third mate contributed to the notion that the spill was a "horrible mistake" and a "freakish accident."[3]

Shivering, oil-soaked otters became the iconic victims, and the ostensibly drunk Hazelwood became the obvious villain in the news story. Yet Exxon eventually became the focal point for popular anger and citizen protest as the coverage moved beyond the drunken captain to implicate corporate rapacity and carelessness. The pathos-inducing pictures of oil-soaked creatures contrasted with the image of Exxon as ruthless despoiler of the Alaskan wilderness.

Still, the visual media reduced the complex dynamics of energy policy and environmentalism to easily understood tropes of blackened beaches and doomed wildlife, and of a formerly pristine landscape ravaged by oil. These pictures encouraged audiences to feel saddened by the devastation and enraged at the company responsible for the spill. Yet feeling upset about oiled otters and angry at Exxon were easy emotions for mass audiences to claim, as they were never asked to consider the larger energy and environmental policies that exacerbated the national dependence upon oil. The imagery repeatedly suggested that Hazelwood and his employer were ultimately responsible for the spill, and thus let a range of other actors—including other major oil companies, the federal government, and viewers themselves—off the hook. While Exxon complained that the coverage preyed on spectator feelings to malign the corporation, many environmentalists believed that the emotional power of oil spill imagery also worked to obscure other environmental crises in the making.

The story of the *Exxon Valdez*—and its emotion-saturated media coverage—provides an iconic example of what the philosopher Slavoj Žižek terms "subjective violence": sudden, immediate bursts of violence typically "performed by a clearly identifiable agent." This highly visible mode of violence, he argues, "is seen as a perturbation of the 'normal,' peaceful state of things." The Alaskan landscape—the so-called last frontier, the putatively pristine wilderness of Prince William Sound—provided the peaceful setting against which the horror of the oil spill would be viewed. The sudden violence of the spill and the destruction visited upon this sublime setting worked to condemn Exxon as spectators expressed moral outrage at the devastation of wildlife and surrounding ecosystems. Yet this focus on the shocking violence in Prince William Sound also represented a failure to grapple with what Žižek refers to as "systemic violence": "the often catastrophic consequences of the smooth functioning of our economic and political systems." Even as scientific theories of global warming gained increasing acceptance, the long-term risks of fossil-fuel dependency, including the slow, systemic violence of climate change, seemed far removed from the wreckage of the *Exxon Valdez*. While some environmentalists warned of these long-term, accretive dangers, dominant views of the spill underwrote the green consumerist strategy of boycotting Exxon, an effort to punish this clearly identifiable agent for its environmental sins. As with the Alar crisis, this campaign presented the market as a promising realm for environmental politics, a way for green consumers to protest the sudden violence of the oil spill.[4]

Five days after the spill began, Roger Caras stood on an island in Prince William Sound. With snow-covered mountains in the distance and gentle waves lapping the rocky shore, the ABC News reporter marveled at the majestic setting but then proceeded to offer viewers a disturbing prophecy. "Up until a week ago, this was a pristine wilderness," Caras announced. "Now, it's a biological time bomb." His report framed the spill as a catastrophe in progress. The visuals showed a variety of creatures, innocent and playful, unaware of the devastation that the spill would unleash. "These are killer whales playing near the grounded tanker," Caras explained. "Even whales, the most intelligent of animals, can't understand what has happened to their world. If the oil gets into their lungs or if they eat enough contaminated fish, it could mean their death."[5]

Caras's segment constituted a roll call of impending doom. He emphasized the scale of potential devastation, and urged viewers to see the oil-soaked creatures as a preview of the monumental casualties that would follow: "The oil-covered birds seen here and doomed to die are a small sampling of what is to come." Following these close shots of individual birds, panoramic images depicted vast quantities of migrating birds. "There is no way of warning them of what is waiting," Caras commented. "Many of them will die." His report closed by looking at sea otters, the species that would soon emerge as the spill's iconic victim. While otters frolic in the water, Caras warned: "Sea otters . . . can't be protected from a contaminated sea. . . . Oil on their fur means certain death." Throughout the segment, Caras played the prophet, foreseeing the death of innocent wildlife.[6]

As they had done after the Three Mile Island accident ten years earlier, TV news used the frameworks of crisis and catastrophe to shape popular perceptions of the *Exxon Valdez* spill. While Three Mile Island never became the worst-case scenario, never devolved into the dreaded China syndrome, the *Exxon Valdez* offered overwhelming evidence of catastrophe. Night after night for almost four weeks, newscasters described the steadily worsening conditions. In these reports, news correspondents, government officials, and environmentalists frequently used apocalyptic language to describe the spill's damage. One "compared it to the devastation of the atomic bomb," while another said it was not only the "nation's biggest oil spill," but "may also be the nation's worst environmental disaster."[7]

Media coverage of the *Exxon Valdez* drew on familiar tropes to frame this crisis. Much as newscasters had done after the Three Mile Island accident, Caras and other correspondents positioned TV spectators as witnesses to potential calamity. Much as had been done after the Santa Barbara oil spill, reporters and image makers offered prophecies of doom by showing otters, birds, and other wildlife just before they perished, and warning of their future death. Unlike in the case of Three Mile Island, though, the media declared this event a catastrophe and circulated visual evidence to verify the scale of death and devastation. To a much greater degree than after the Santa Barbara spill, the coverage of the *Exxon Valdez* accident not only used the camera's prophetic capacity, not only foretold of the imminent loss of wildlife, but also broadcast actual images of dead animals. These documents of death began beaming into American homes a few days after the spill began, and continued to air for weeks. Like Dawson's otter photograph, these depictions of particular animals that were at risk or already dead elicited feelings of sadness and encouraged audiences to worry about the fate of vulnerable creatures. Such pictures invited an emotional response, focused closely on specific victims—and on the oil coating their feathers and fur—to reveal how the spill assaulted and ultimately annihilated defenseless beings.[8]

Television news not only portrayed individual deaths but sometimes showed the piles of dead bodies collected and tabulated by government scientists. One correspondent described the growing heaps of garbage bags filled with animals as "scenes of the grisliest work of all—a biologist who used to count the living now counting the dead." Standing next to a stack of specimens, the biologist explained that he had not become "numb to it," had not become completely desensitized to the sickening sight, "but you can't let your emotions run ahead of you or you couldn't do the job." As he completed this inventory of death, the scientist could not fall prey to his emotions, could not be overwhelmed by grief and sadness. Yet his comment implicitly acknowledged that such an emotive response would be expected of viewing audiences who did not have to spend their time, day after day, grimly tallying the losses.[9]

Media coverage of previous oil spills, especially the San Francisco Bay spill of 1971, had emphasized the story of altruistic volunteers trying to save helpless creatures. Coverage of the *Exxon Valdez* spill followed this pattern and also generated images that singularized the moment of rescue. Taken by AP photographer Jack Smith, one of the most frequently circulated images showed an oil-soaked bird, wrapped in black plastic and cradled in human hands (fig. 14.2). The image reveals

FIGURE 14.2. Oil-covered bird in Prince William Sound, Alaska. Photograph by Jack Smith, 1989. AP Photo / Jack Smith.

the victim's sentience and apparent alarm. Photographed in profile, the bird appears, like Dawson's otter, to emit a cry for help and to produce anthropomorphic empathy; its eye returns the spectator's gaze, while its wide-open mouth signifies suffering and offers a desperate appeal for assistance. By capturing this particular moment, the photograph isolates the individual bird and the individual volunteer from the broader context of the oil spill. The photograph does not promise that the bird will survive, nor does it guarantee the efficacy of individual action. Instead, the image holds the moment of rescue forever in suspension, leaving the viewer to ponder the uncertain fate of oiled wildlife.

Similar scenes of blanket-wrapped birds and animals appeared across a wide range of media. Such reports could have worked to obscure the scale of devastation, and could have provided touching, even uplifting stories of individual animals restored to health. Yet many TV news broadcasts noted instead that this "valiant effort was proving futile." The rescue workers, portrayed as environmental altruists, people who desperately wanted to stem the tide of devastation, nonetheless faced a crisis of such magnitude that they could only barely restrain the rising death toll. Close-up shots of otters in cages emphasized their vulnerability and helplessness, and invited audiences to respond emotionally to their oil-induced torment. Yet for every story of an otter saved—and of the occasional creature sent away to Sea World—newscasters would inform viewers that at least half of the rescued animals would eventually die, and that multitudes more would perish all across the region.[10]

Media images also emphasized the sublime setting of Prince William Sound—the glacial mountains, remote islands, and rocky fjords—and encouraged audiences to feel that a pristine wilderness was being ravaged by the spill. TV news and popular magazines used familiar visual codes—especially the sublime spectacle of snow-capped mountains and the vast populations of shorebirds and marine animals—to signify the region's inspiring beauty and life-sustaining power. The wilderness aesthetic conveyed the image of a supposedly pure setting suddenly marred by the violence of the oil spill. While close-up shots of oiled otters and birds evoked sympathy for particular victims, panoramic views of the surrounding landscape elicited feelings of awe and wonder for the Alaskan wilderness now under siege.[11]

On the same night that TV news revealed the results of Hazelwood's blood alcohol test, they also all aired a statement by Paul Yost, the commandant of the Coast Guard. At a press conference, Yost mocked the crew's ineptitude by describing "the area they went aground" as "not treacherous" and "ten miles wide." "Your children could drive a tanker up through it," he claimed. For a while, the drunken captain and the incompetent crew provided the media with an easily identifiable culprit, and made the disaster appear as an accident caused by booze and blundering. Hazelwood became, as *Time* put it, "America's Environmental Enemy No. 1." The soused skipper became an icon of ecological irresponsibility and the butt of late-night TV show jokes. In a statement played back on all major TV news broadcasts, a judge accused the careless captain of acting as the "architect of American tragedy." "We have a . . . man-made destruction that probably has not been equaled since Hiroshima," the judge announced. Meanwhile, on *Late Night with David Letterman*, the popular TV comedian listed Hazelwood's top ten excuses for the spill, including this alcohol-themed zinger: "I was just trying to scrape some ice off the reef for my margarita."[12]

Yet, even as Exxon tried to shift the blame onto Hazelwood, the company's response to the spill—widely viewed as botched and bumbling, marked by delays and hampered by inadequate supplies and staff—led many to condemn its cleanup operation and hold the corporation primarily responsible for the catastrophe. As public outcry mounted, Exxon tried to restore its image by purchasing full-page ad space in major newspapers and magazines. Like the Union Oil ad that appeared after the Santa Barbara spill, the Exxon ad featured an open letter from the company chairman. But many Americans rejected Exxon's image-restoration strategy. Denouncing the ad, one *Newsweek* reader explained: "Exxon's letter of apology that appeared in America's newspapers and magazines is an insult to the public. The Alaskan oil spill was a national disaster; the cause was a disgrace. . . . Exxon, save the advertising dollars. This spill is more than just an image problem."[13]

The Exxon name and logo soon took on sinister meanings. Both had been developed by Raymond Loewy, the twentieth century's most influential industrial designer, who had been hired by Standard Oil of New Jersey in 1966 to change its brand name from Esso. After settling on the name Exxon, Loewy decided to maintain the red lettering and blue border from the old logo, but made several changes to the design, including, most of all, calling attention to the new company name by "placing the visual emphasis on the double x's." Loewy explained that the intersecting diagonals infused the *x*'s with dynamism and fulfilled his goal of creating "a very high index of visual memory retention. . . . In other words, we want anyone who has seen the logotype, even fleetingly, to never forget it."[14]

In the aftermath of the spill, the Exxon logo would certainly not be forgotten, but now it would signify dead otters and defiled wilderness and would remind viewers of the company's environmental destruction. Some network news broadcasts would use the Exxon logo as part of the backdrop image at the beginning of oil spill stories, thus reinforcing the corporation's culpability for the disaster. In protest rallies covered by the media, some environmentalists altered Loewy's interlinked *x*'s by turning the letters into swastikas. Editorial cartoonists routinely skewered the corporation for the spill and for grossly exaggerating the success of its cleanup operation. Mike Peters, for example, rejected Exxon's oil-removal claims as patently false. In his syndicated newspaper cartoon, a young girl stands in her obviously disheveled bedroom. She looks up at her at mother and tries to explain away the messy conditions. "But I *did* clean my room . . . at least by Exxon's standards."[15]

As the Exxon brand became tainted, a number of environmental groups urged consumers to boycott the company, to cut up their Exxon credit cards and no longer patronize Exxon service stations. One boycott-promoting advertisement featured a photograph of a dead otter and began: "This sea otter can't fight back, but you can. Boycott Exxon" (fig. 14.3). The ad appealed to readers as emotional beings, as citizens who felt a sense of revulsion at the spill's devastation and moved by the images of oiled wildlife. In contrast to this affective public, to the spectators who felt sympathy and compassion for nature, Exxon appeared as a cruel, emotionless entity. The boycott aimed to punish Exxon "by sending them a reminder that'll hurt in the only place most big corporations have any feelings: Their wallet." By generating these Manichean divisions between innocent nature and vile Exxon, between emotive citizens and soulless corporation, boycott ads ennobled and empowered participants, made consumers feel that their decision to fill up at other stations would be a meaningful political gesture, one that would exact retribution on Exxon and help protect the environment.[16]

Many boycotters relished the opportunity to castigate the company. As one explained, "What am I doing to express my disgust and disappointment? I am boycotting Exxon gas stations and returning my torn-up credit card to Exxon chairman L. G. Rawl!" Another wrote to the company: "Here's my Exxon card and Exxon travel card: stick them up your oil spill!"[17]

The boycott made the market appear as the prime arena of environmental action, the ultimate arbiter of sustainable behavior. According

THIS SEA OTTER CAN'T FIGHT BACK, BUT YOU CAN. BOYCOTT EXXON.

You've heard plenty about the millions of dead fish and birds and sea otters in Prince William Sound. You know that the gooey residue of this act of negligence will haunt us for years. You know Exxon has done precious little to clean up the mess.

These are the horrors we face today, but what about tomorrow? Once the furor has died down, will Exxon and other corporations have learned from this tragedy? Or will they go back to business as usual?

We can make sure that won't happen by sending them a reminder that'll hurt in the only place most big corporations have any feelings: Their wallet.

The message we want to deliver is this: Environmental responsibility is as important as the quality or price of a company's products. We want all companies to know that from now on people are going to be considering their environmental reputation when they buy.

What can you do? First, make lots of copies of this ad. Take them down to your local stores, schools, offices, churches—wherever people gather.

Then, fill out and send the coupon yourself.

Above all, register your protest with your steering wheel, by turning away from Exxon gas stations.

Hopefully, we've all learned a lesson from this catastrophe. Every one of us needs to take personal responsibility for the preservation of this fragile planet. Here's where we can start.

CITIZENS FOR ENVIRONMENTAL RESPONSIBILITY

We are a non-profit group organized to help educate people about corporate environmental responsibility. We have no agenda other than preserving a healthy planet. You can help by freely reproducing or faxing this coupon and making it a huge chain letter. For other ideas on how to help, please call 408-399-0339.

We want to make sure your response is accurately counted. Certainly, Exxon won't tell us how many coupons they get. Please send them to us and we'll forward them to Exxon.

TO LAWRENCE G. RAWL, Chairman, Exxon Corporation
**c/o Citizens For Environmental Responsibility
101 Church Street, Suite 6, Los Gatos, CA 95032**

Sir: In response to Exxon's reckless disregard of its obligation to protect the environment of Prince William Sound, Alaska:

☐ Enclosed is my Exxon credit card cut in half.

☐ I don't have a credit card, so this is my pledge not to buy Exxon products for at least all of 1989.

Signature _____

Address
(optional) _____

FIGURE 14.3. Citizens for Environmental Responsibility advertisement, 1989.

to the logic of the boycott, economic actors, free to buy gas wherever they wanted, could choose to avoid the Exxon logo, and could express their environmental concern by rewarding other corporations with their purchases instead. "Above all," the boycott ad announced, "register your protest with your steering wheel, by turning away from Exxon gas stations." In this framing of environmental politics, the structural dimensions of energy policy receded behind appeals to virtuous consumption. The market became the perpetual voting booth of consumer culture, the realm where freely acting individuals could, as *New York Times* columnist Thomas Friedman put it, "vote every hour, every day" through their purchasing decisions. The boycott fostered an image of environmental citizenship as an enlightened form of consumerism. This activism began in the domain of individual choice, with the driver who maximized ecological values by buying gas elsewhere. "Hopefully, we've all learned a lesson from this catastrophe," the ad concluded. "Every one of us needs to take personal responsibility for the preservation of this fragile planet. Here's where we can start."[18]

By trafficking in scenes of ecological devastation, the visual media critiqued Exxon and seemed to align itself with the environmental cause. Indeed, following a pattern set by the conservative and industry reaction to *The China Syndrome*, Three Mile Island, and the Alar scare, Exxon officials claimed that the media's emotion-laden imagery unfairly discredited the corporation. In an interview with *Time*, company chairman Lawrence Rawl equated mass-media imagery with environmental protest. According to him, both the media and environmentalists deployed powerful images to exploit audience emotions and promulgate anti-Exxon propaganda. "This tanker went on the rocks," Rawl explained, "and visually it was perfect for TV and . . . for pictures of oily birds in the printed media. How would those environmentalists ever let that go?" Another Exxon official charged: "The public's reaction is totally irrational." Company officials maintained that the media used imagery to venerate nature and vilify Exxon, to manipulate public feelings and malign the corporation.[19]

Six months after the spill, however, many mainstream media sources began to reframe the event, claiming that it was not so catastrophic after all. In the most blatant example, *U.S. News & World Report* ran a cover story titled "The Disaster That Wasn't." The cover image featured a man, outfitted in helmet and work gloves, directing a high-pressure hose at oil-stained rocks along a blackened beach. While the image could be read as a sign of humanity's ability to use technology to cleanse the natural world, it circulated as part of a broader set of media images that questioned the effectiveness of Exxon's cleanup effort. Indeed, many media reports mocked the use of high-tech hoses as well as the sight of "hundreds of men and women . . . wiping oil off rocks by hand," using "special absorbent pads" to clean them "one by one." Rather than continuing to endorse this futile effort, *U.S. News, Newsweek*, and other popular sources invoked the power of nature to heal itself. By celebrating natural resilience, these reports sought to counter what they described as "one-dimensional television images" that had depicted Prince William Sound as "an ecological disaster zone." While initial coverage had mourned the tragic spectacle of doomed wildlife, the popular media now fostered hope in the potential for natural recovery.[20]

In a January 1990 cover story, *National Geographic* likewise evoked the theme of nature's healing powers. The cover photograph by Natalie Fobes depicted an oiled pigeon guillemot, nestled in black plastic, its eye hauntingly gazing out (fig. 14.4). Like Jack Smith's photograph of an oil-soaked bird, this image isolates a single creature at the moment of rescue. This photograph, though, does not reveal the nurturing presence of human hands. Hovering above a blurred landscape, the guillemot's beak guides the reader to the cover story's key question: "Alaska's Big Spill: Can the Wilderness Heal?" While *National Geographic* had presented the 1969 Santa Barbara oil spill as a symbol of a much broader, all-encompassing environmental crisis (see fig. 2.3), this cover story instead emphasized the effects of the spill on the Alaskan wilderness and appealed to nature's restorative powers. "Sooner or later," the story concluded, "through human error or simply through the perils of the sea, spilled oil will assault another shore. And sooner or later, the damage will have to be left to nature to repair."[21]

This focus on nature's capacity to heal itself became part of Exxon's image restoration strategy. Exxon and the popular media sought to turn public attention away from the sudden violence of the oil spill by emphasizing instead the long-term prospects of natural recovery. This strategy has since been deployed by the oil industry following other catastrophes, including the 2010 *Deepwater Horizon* disaster in the Gulf of Mexico. In this latter case, BP and the mainstream media rushed to declare the disaster over as soon as the wellhead was capped. In both cases, the media's emphasis on the most visible evidence of destruction obscured the persistent presence of petroleum hydrocarbons and their long-term, cumulative damage to marine ecosystems.[22]

FIGURE 14.4. *National Geographic* magazine cover, January 1990. Photograph by Natalie B. Fobes / National Geographic Creative.

Still, the *Exxon Valdez* spill weakened, at least temporarily, the credibility of the oil industry and encouraged environmental reform. In particular, the disaster led to congressional passage of the Oil Pollution Act of 1990. Signed by President George H. W. Bush, this legislation increased penalties and liability for oil spills and allocated more funding for future cleanup operations. Moreover, while Bush followed Reagan in continuing to push for oil drilling in the Arctic National Wildlife Refuge, the *Exxon Valdez* disaster made this position politically untenable, at least for a while.[23]

Yet not every environmentalist believed that the media's intense focus on the spill worked to advance their cause. Much like Žižek's analysis of subjective versus systemic violence, their critiques contrasted the visual horror of the *Exxon Valdez* disaster with long-term, less-visibly-spectacular dangers. The radical environmental thinker Murray Bookchin observed: "We tend to think of environmental catastrophes—such as the recent *Exxon Valdez* oil-spill disaster . . .–as 'accidents': isolated phenomena that erupt without notice or warning. But when does the word *accident* become inappropriate? When are such occurrences inevitable rather than accidental?" Bookchin claimed that media temporality and the camera's relentless focus on the most visible signs of degradation made it difficult for audiences to glimpse the structural causes of environmental danger. Bookchin used a visual metaphor to explain how the popular media packaged ecological crises for public consumption. In these depictions, he concluded, "Our problems . . . are episodic rather than systemic; the scene dissolves, the camera moves on."[24]

The environmental writer Bill McKibben, whose first book *The End of Nature* was to appear later that year, also challenged conventional views of the spill. McKibben wanted audiences to recognize what was being left out of coverage that emphasized the spill's most obvious effects: its annihilation of wildlife and marring of the sound. Writing on the *New York Times* op-ed page, he began by describing the widely shared feelings of spectators exposed to the horrific imagery: "The pictures of the Alaskan spill sickened all who saw them." He immediately shifted, though, from the emotionality of oil spill imagery to a thought experiment about what would have happened to all that oil had the supertanker not run aground. "But say that Capt. Joseph J. Hazelwood had steered his ship straight—say that his oil had reached its destination safely and had been burned in power plants and car engines," McKibben wrote. "The 10 million gallons of oil . . . would have released into the atmosphere about 60 million pounds of carbon in the form of carbon dioxide. It is the buildup of this carbon dioxide, more than any other factor, that scientists say will produce the greenhouse effect, raising the world's average temperature three to nine degrees in the next few generations."[25]

McKibben thus reflected on the failure of the visual media to communicate evidence of climate change—a gradually escalating crisis rather than a sudden episode easily depicted by images and easily digested by spectators. The media's focus on the pristine wilderness under siege naturalized the long-term systemic violence of fossil-fuel dependency. "The greenhouse effect . . . is not the result of something going wrong," McKibben continued. "It doesn't stem from drunken sailors, inadequate emergency planning or a reef in the wrong place. It's harder to deal with than that because it's just a result of normal life." While the visual media proved adept at publicizing dramatic examples of environmental disaster, the "normal," systemic assault on the ecosphere often remained hidden from view. "When you look at pictures from Alaska," McKibben concluded, "remember this: The ship that is our planet has a gaping hole in its side, too, and carbon dioxide is pouring out." McKibben recognized how media coverage of the *Exxon Valdez* obscured the accretive dangers of petro-dependency: Pictures of oiled otters, no matter how heart-wrenching they might be, could not convey the hazards of global warming, could not depict the catastrophic consequences of slow violence.[26]

At the end of 1989, *Time* magazine selected Tony Dawson's otter photograph as one of the "most amazing pictures" of the year and the quintessential oil spill image, the one picture that most powerfully captured the meanings of the *Exxon Valdez* disaster. Indeed, Dawson's photograph became the only oil spill image to appear in *Time*'s retrospective. It must have been particularly striking for audiences to view this photograph alongside the other images chosen by *Time* to encapsulate 1989—a "revolutionary year," the magazine explained, marked by "visions of freedom, defeat and resolve." *Time* hailed the iconic photograph of the lone protester facing a column of tanks near Tiananmen Square, "a Chinese man standing down a tyrannical regime," as "the most extraordinary image of the year. . . . It is flesh against steel, mortality against the onrush of terror, the very stuff of courage." The special gallery also featured a photograph of the Berlin Wall being dismantled, an epochal event that signified the collapse of the Iron Curtain and the end of the Cold War. "With all the righteousness of a biblical drama," the magazine noted, "East Germans march and march against their Communist masters until, astonishingly, the Wall comes tumbling down." From Beijing to East Berlin, *Time* celebrated the human quest for freedom and the struggle to overcome tyranny around the globe. While the former Soviet bloc lurched toward freedom, while Chinese protesters struggled valiantly, if unsuccessfully, to challenge their totalitarian regime, the otter, defenseless and devoid of agency, had succumbed to the toxic spillage of the *Exxon Valdez*.[27]

As the Cold War came to an end, and as Earth Day 1990 approached, the sense of global consciousness became more central to popular framings of environmentalism. If the Berlin Wall could come down, if one protester could steadfastly confront a column of tanks, then perhaps Americans and people around the world could come together to face common ecological problems and foster hope for the future. Even as these portrayals of the movement imagined change at the global level, they often ignored power inequities between different peoples and nations, and tended to zoom in on the individual as the ultimate agent of ecological transformation.

Like other *Exxon Valdez* images, Dawson's otter may have caused viewers to despair about the environmental future and to see the spill as symbolic of a broader environmental crisis. Earth Day 1990 would directly address this emotional state and would seek to empower individuals by emphasizing personal action and green consumerism. Although some mainstream groups embraced this rhetoric of individual responsibility, other activists wondered whether it threatened to turn environmentalism into a depoliticized form of green therapy.

FIFTEEN

Global Crisis, Green Consumers: The Media Packaging of Earth Day 1990

Meryl Streep walks into a bar. She appears flustered. She exhales loudly and then exclaims, "I need a drink. I need a drink please."

The bartender, played by Kevin Costner, grabs a can of beer, pours it into a glass, and asks, "You're OK?"

"No, I'm not OK," Streep replies. "I couldn't stand it out there. I hate thinking about this stuff. What if we have screwed it up so bad that they can't fix it? I mean, it's crazy what we're doing."

Streep has just come from an Earth Day 1990 rally in which speaker after speaker has bemoaned the severity of the environmental crisis. Costner agrees with his anxious customer that the situation is "really bad," even "terrifying."

She chides him, "I thought bartenders are supposed to make you feel better."

"I can do that," he claims. Costner then describes everyday actions, most notably recycling, that individuals can take to demonstrate their environmental concern. When Streep questions whether these "tiny" and "pathetic" gestures can ward off the looming crisis, he deflects her pessimism by emphasizing the therapeutic value of recycling. "You need to do something instead of being afraid," he

239

says. "You do something, you're gonna start to feel better about your-self." He hands her the empty beer can and points to a recycling bin behind the bar. As she prepares to aim and shoot, he advises: "Take your time. Don't rush this. This could change your life" (fig. 15.1).

This fictional barroom exchange occurred near the end of *The Earth Day Special*, a star-studded TV program that ABC aired during prime time on April 22, 1990, the twentieth anniversary of Earth Day. The sequence encapsulated the emotional dynamic that ran throughout the show—Streep's anxiety soothed by Costner's effort to make her "feel better." The entire program, much like actual Earth Day events that took place that day, stressed the personal and psychological di-mensions of environmentalism. *The Earth Day Special* encouraged mass audiences to move beyond emotional paralysis by assuming personal responsibility for environmental problems. Similarly, Earth Day 1990 encouraged Americans to find therapeutic relief through recycling, tree planting, and other forms of individual action.[1]

The *Earth Day Special* provides a particularly revealing example of the media embrace of environmentalism. Hailed as the largest gather-ing of entertainers ever to appear on a single TV program, the show featured an astonishing array of celebrities from diverse entertainment

FIGURE 15.1. Meryl Streep and Kevin Costner in *The Earth Day Special*. Frame capture from television program, 1990.

genres. Bette Midler starred as "Mother Earth," bedecked in a costume "made of real Los Angeles trash" and suffering from the escalating effects of the environmental crisis. Neil Patrick Harris, the star of *Doogie Howser, M.D.*, reprised his sitcom role as a teenage doctor, here trying to revive Mother Earth. Meanwhile, the casts of *The Cosby Show*, *Cheers*, and *The Golden Girls* all contemplated how they could help save the planet. An impressive roster of movie stars—including Morgan Freeman, Dustin Hoffman, and Robin Williams—performed in various skits, while *The China Syndrome*'s Jane Fonda, Michael Douglas, and Jack Lemmon all expressed their ongoing environmental concern. *The Earth Day Special* also featured legendary music producer Quincy Jones along with an entourage of popular rappers, including the Fresh Prince, Tone Loc, and Queen Latifah. In addition to the extensive list of human celebrities, Bugs Bunny, Kermit the Frog, and the adorable alien E.T. appeared on the program.[2]

It would be easy to dismiss *The Earth Day Special* as an entertaining spectacle or empty sideshow, a meaningless distraction from the serious events and politics that constituted Earth Day 1990. Sure, the notion of Bette Midler appearing as an ailing Mother Earth or of E.T. offering ecological wisdom to human beings would make an easy target for ridicule. But what makes *The Earth Day Special* so telling as a cultural text is how it merged so seamlessly with the larger media extravaganza of that day. Across a wide variety of visual and cultural fields, spectators were repeatedly urged to seek both personal therapy and a sense of political empowerment through individual action.

Indeed, Costner's advice to Streep echoed the official motto of Earth Day 1990: "Who says you can't change the world?" Developed by Pacy Markman—the advertising copywriter who created the hugely successful Miller Lite slogan, "Everything you always wanted in a beer, and less"—this tag line appeared on Earth Day T-shirts, bumper stickers, and posters, and captured the emotional politics of environmentalism in this period. Posed as a question, the slogan nevertheless offered the implicit promise of political agency, the empowering sense that individuals could make a difference.[3]

As we have seen, the discourse of individual responsibility has long played a crucial—and contested—role in popular environmentalism. During the time surrounding the first Earth Day celebration in 1970, environmentalists had encouraged individuals to live in a more ecologically sustainable fashion. Yet many of these same activists rejected the mass media's exaggerated emphasis on individual action and worried that Pogo, the Crying Indian, and other appeals to personal responsibility obscured power relations and shifted attention away from the corporate and government structures that had led to environmental degradation. Likewise, throughout the 1970s, many activists continued to question the individualist frame and sought to imagine more far-reaching collective solutions to the nation's energy crises.

What made the twentieth anniversary of Earth Day so anomalous was the extent to which its leading organizers now embraced and promoted personal responsibility for the environmental crisis, the same discursive frame they had repeatedly challenged. For instance, Denis Hayes, the national coordinator for both Earth Day in 1970 and Sun Day in 1978, also chaired Earth Day 1990, but this time around, he repeatedly personalized the environmental cause. Similarly, Gaye Soroka, who had participated in Earth Day 1970 and helped plan Earth Day 1990 activities in San Diego, dismissed the event's first incarnation for only prescribing "legislative action without stressing how individuals can make a difference." "We had a giant rally," she recalled, "we marched and listened to political speakers and bands, lay in the sun and drank a lot of beer. We signed petitions and then went home. It didn't get anywhere because it wasn't personalized."[4]

Just as Costner tried to make Streep feel better by showing her where to toss a beer can, the media spectacle of Earth Day packaged environmentalism as a form of virtuous consumption. From the popular guidebooks listing what you could do to save the planet to the endorsement of green consumerism by Earth Day organizers, lifestyle decisions and the marketplace became the dominant sites of environmental activity. During this pivotal moment, mainstream depictions of the movement worked to enshrine neoliberal models of environmental citizenship.[5]

This chapter will view Earth Day 1990 through three broad contextual lenses—the dramatic expansion of recycling and green consumerism, the media's marginalization of the environmental justice movement, and the end of the Cold War—to understand how popular environmentalism intersected with the emotional life of capitalism. According to media imagery, recycling and green consumerism provided personal empowerment to the consumer citizen. Environmental anxiety thus became incorporated into the market, and political action became synonymous with individual buying choices. In contrast, the burgeoning environmental justice movement—organized primarily by low-income grassroots activists of color—challenged market imperatives by resisting corporate efforts to burden minority communities with toxic waste, and by calling for radical reductions in the production of environmental pollutants. Even as Earth Day 1990 sought to rebrand environmentalism as a multicultural cause and thereby move beyond the movement's lily-white image, the systemic critiques of environmental justice activists did not mesh with the dominant frames of personal empowerment and individualized politics. Meanwhile, the end of the Cold War reinforced the increasingly global vision of environmental concern, but also underwrote the neoliberal faith in the marketplace. Even as capitalism distributed environmental risk unevenly, both within the United States and on a global scale, the market appeared to offer a sustainable path out of the abyss.[6]

Most media portrayals of Earth Day 1990 accepted certain long-term, escalating problems as a given, as the frightening reality that led Streep and others to articulate their environmental anxiety. Nevertheless, short-term, consumerist solutions helped soothe this anxiety. By framing structural problems in emotional terms, this therapeutic mode obscured environmental inequities and ultimately displaced the long-term sense of danger. Recycling and other forms of individual action accomplished something meaningful for consumers, making them feel that they were participating in a vital project to save the planet. The media offered ample evidence of capitalism's destruction of the ecosphere, but also prescribed emotionally satisfying ways for citizens to govern at a distance and gain feelings of environmental hope.

In January 1990, ABC News on two successive evenings promoted green consumerism and individual environmental action. John Javna, author of *50 Simple Things You Can Do to Save the Earth*, which became the bestselling title in the genre of do-it-yourself environmentalism, appeared one evening to discuss the book's underlying principles. "We're faced with environmental problems that seem so overwhelming to us," Javna observed. Like *The Earth Day Special*, Javna wanted to move beyond doomsday scenarios that left individuals feeling powerless, and to show viewers how everyday actions could both help the environment and make them feel better. "This doesn't have to be ponderous, terrifying, and difficult," Javna concluded. "It is really satisfying."[7]

In another broadcast ABC News explicitly linked individual action to consumerism by portraying environmentalism as an enlightened form of shopping. Once again, the report stressed how Americans felt "frustrated" by the magnitude of environmental problems, but then it sought to counter these feelings of powerlessness by revealing how, "on a personal level, there's a lot they can do." After describing an array of "green" items now appearing in supermarkets and shopping malls and lauding McDonald's for introducing a recycling program, the correspondent concluded on an upbeat note: "Consumers have increasing strength to make business do the right thing. They have the most powerful weapon—their wallet."[8]

Soon after Earth Day, Denis Hayes helped launch Green Seal, an organization that would "guide people in being 'environmentally sensitive' consumers." Products that met certain criteria would receive the Green Seal logo—a green check mark emblazoned over a blue globe—to signal their ecological virtue. Hayes argued that consumer choice and the market promised to achieve what protest movements and government policies could not. "The world is in markedly worse shape than it was twenty years ago," he explained. "It's a source of incredible frustration and leads to the conclusion that past strategies are inadequate, that exclusive focus on the government, on trying to influence the levers of power, is important but not sufficient." Rather than relying so much on the government, environmentalists should now place their faith in the market and cultivate the "critically important roles that people have to play acting as investors and consumers." If governments could not protect the citizenry from environmental risks, then perhaps market-based solutions and personal actions could simultaneously save the planet and soothe the individual.[9]

This emotional dynamic—the sense of crisis and the longing for individual solutions—provided the impetus for *The Earth Day Special*. The show's executive producer, Armyan Bernstein, recalled the experience of "watching news reports" with his girlfriend during the summer of 1988 as frightening footage of "hypodermic needles, vials of blood and other medical refuse washing up on East Coast beaches" filled the screen. "My girlfriend sat there and started to cry," Bernstein said. "I asked her what was wrong, and she said, 'I just feel so bad for the Earth.'" Although he and his girlfriend thought about writing a save-the-planet manual, Bernstein believed that a TV special could more effectively raise awareness and "change people's consciences."[10]

The Earth Day Special began in the private sphere, in the living room of a married couple played by real-life husband and wife Danny DeVito and Rhea Perlman. Sitting on their couch, the fictional Vic and Paula engaged in a conversation in which she evinced environmental concern while he appeared completely clueless. "Do you know what today is?" Paula asked. Vic looked worried, as if wondering whether he had forgotten something important. "Valentine's Day?" With an incredulous look he then asked, "Your birthday?" After Paula informs him that it was Earth Day, he responded, "What the hell is Earth Day?" Paula replied that she had been "reading a lot about the planet" and tried to convince him to watch *The Earth Day Special* with her. He finally conceded to do so, but just for "a couple minutes."

As they watched the television, Vic occasionally picked up the remote to change the channel but found that he could not escape the distressing imagery of the environmental crisis. Even as he switched to game shows like *Jeopardy* and *The New Dating Game*, everyone on his TV seemed obsessed with the environment. At first Vic appeared to feel assaulted by the environmental barrage, but about halfway through the broadcast he stopped channel surfing and became riveted by the Earth Day show.

The Earth Day Special shifted back and forth between Paula and Vic's living room and a town square in which citizens were gathering for an Earth Day celebration. Suddenly, Mother Earth, played by Midler, appeared on the scene. "I'm sick," she told the crowd, "and it's all your fault." After "Mother Earth" was rushed to the hospital, a TV news reporter played by Candice Bergen, the star of *Murphy Brown*, appeared and announced, "There are unconfirmed reports that Mother Earth is dying."

Throughout the show, the town square provided a site where a meaningful public culture could form, and where citizens could gather to discuss environmental problems. At first *The Earth Day Special* seemed to contrast this vibrant public culture with the detached, private experience of spectators watching TV at home. As the storyline developed, though, the program focused increasingly on the private sphere. By the end of the show, the domestic realm came to be presented as both the prime cause of and potential solution to the environmental crisis, displacing the town square as the main site where environmental publics could form and where environmental change could happen.

The Earth Day Special emphasized gradually escalating environmental dangers—including air and water pollution, garbage overload, deforestation, and global warming—to convey the extensive timescale of environmental crisis. Rather than merely focus on the sudden violence of the *Exxon Valdez* and other spectacular disasters, the program, like *Time*'s "Endangered Earth" cover story, depicted the accretive, long-term risks of the global environmental crisis.

This emphasis on long-term danger was reinforced through the program's repeated use of children as emotional emblems of the future. In one scene, three kids brought saplings to a hospital, hoping to give them to Mother Earth to help "slow down global warming." (Tree-planting children were also appearing repeatedly in media coverage of Earth Day 1990 events. Like recycling, tree planting would be celebrated as an emotionally satisfying form of individual action.) In another scene in the program, four children discussed their feelings of powerlessness in the face of worsening environmental conditions. "My mom and I got so sad," one explained. "Things are so messed up. We didn't know what to do about it." The movie alien E.T., with his long, magical finger aglow, then appeared and offered them an enormous book titled *A Practical Guide to How You Can Save the Earth*. E.T. said slowly as his heart began to shine: "For you. For you, people of Earth."[11]

Bergen, as the newscaster, then reported on the emotional transformation felt by the crowd at the town square: "Moments ago, fear gave way to hope, as children took to the stage with a glowing book. People are now taking turns reading from that book." Produced by an innocent, kind-hearted alien and presented to the public by innocent, well-meaning children, the save-the-earth manual resembled Javna's *50 Simple Things You Can to Do to Save the Earth*. Yet the focus on childhood agency, the depiction of children—and E.T.—as the bearers of ecological wisdom, deflected attention from power relations. By turning to children, people who could not yet act as citizens, and upholding them as exemplars of environmental citizenship, *The Earth Day Special* sentimentalized and infantilized popular environmentalism.[12]

Speaker after speaker addressed the crowd in the show, almost all of them emphasizing individual actions to help save Mother Earth. "What we keep hearing over and over tonight is the importance of recycling," Bergen's character observed. The relentless focus on recycling, on green consumerism, and on children as models of environmental citizenship all worked in the program to restrict ecological concern to the home and private life. As the literary critic Stacy Alaimo observes, the narrative conceit of *The Earth Day Special*—that Mother Earth blamed her "children" for the environmental crisis and now sought their aid—depicted pollution as being caused by "unruly children, thus making the problem personal and familial instead of political and systemic. It shifts the focus from capitalism to the home and places the blame . . . ultimately on homemakers."[13]

Near the end of *The Earth Day Special* Vic asked, "What we are going to tell our grandchildren?" Paula reminded him that they did not yet have grandchildren, but Vic looked beyond the present to contemplate the accumulating dangers of the environmental crisis. He wondered aloud if his grandchildren might one day ask: "Did we see this coming? Did we know that this would happen? What are we going to tell them? That we didn't care enough?" His questions recalled the Ad Council's

1971 Keep America Beautiful ad that pictured a young girl asking her father, "Daddy, what did you do in the war against pollution?" (see fig. 5.3). The character Vic, anticipating these future questions from his future grandchildren, experienced an eco-conversion and said that he now wanted to engage in everyday actions to save the planet.

During the period surrounding Earth Day 1990, recycling became the most popular and prescribed form of environmental action. It helped consumers feel like virtuous ecological subjects and offered what many described as much-needed hope. The evening before Earth Day, CBS News profiled Naperville, Illinois, an affluent city that had become a national leader in recycling. Interviewed at home, one white woman described why recycling had become so meaningful to her: "It does feel good," she commented. "You feel like you're helping." Recycling provided emotional rewards and feelings of affective empowerment that countered the doomsday scenarios of environmental crisis.[14]

As a ritual to assuage environmental anxiety, recycling performed a therapeutic function for individual consumers. According to the historian Jackson Lears, the therapeutic ethos of US consumer culture often places "emphasis on self-realization through emotional fulfillment" and leads to "the devaluation of public life." Other critics of consumer culture, most notably Christopher Lasch, have condemned consumerist therapy for its selfishness and narcissism, for promoting short-term fixation on the individual self rather than long-term consideration of public life and the environmental future. Recycling, though, would be hailed for its supposed power to fuse the personal with the political and the planetary. Rather than elevating personal gratification over the public good, recycling seemed to provide a way for consumers to merge an inner-directed sense of well-being with an outward-looking concern about the environmental crisis. The recycling logo signified this desire to make consumer culture more sustainable by enlisting individuals in the environmental cause.[15]

The plastics industry proved particularly adept at manipulating consumer desire for environmental change, and at exploiting images of sustainability to shore up an unsustainable agenda. More than ever before, the recycling logo would be pressed into the service of maintaining dominant production practices. Indeed, during the late 1980s, the plastics industry began to alter the original logo by placing numerals representing different grades of plastic in the center of the symbol (fig. 15.2). "Plastic packaging bearing the triangular symbol misleadingly telegraphed to the voting consumer that these containers were

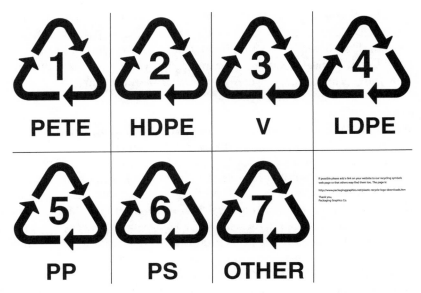

FIGURE 15.2. Recycling logos with resin identification codes, developed in 1988 by the Society of the Plastics Industry.

recyclable and perhaps had even been manufactured with reprocessed materials themselves," the writer Heather Rogers argues. "But often neither was the case." This icon of environmentalism thus helped legitimate the continual expansion of plastics production. Even though the rate of plastics recycling never came close to keeping pace with manufacture of new plastics, even though the industry frequently opposed measures to expand recycling or regulate its environmental practices, the recycling logo signified sustainability and affective empowerment. "As the 1990s wore on," Rogers concludes, "the chasing arrows and printed appeals to 'please recycle' adorned every type of bottle, jar, can, envelope and wrapper, as if they were endorsed by Mother Nature herself. . . . Regardless of industry's actions, the rhetoric of recycling targeted individual behavior as the key to the garbage problem, steering public debate away from regulations on production." As a visual icon of environmental hope, the recycling logo reassured consumers that the three chasing arrows would cycle on and on, that the Escher-inspired loop would promise ecological permanence.[16]

While recycling promoted the notion of governing at a distance, of citizens taking individual responsibility for the environmental future, this neoliberal solution depended upon massive government support, from new recycling laws to the funding of recycling programs. Between 1989 and 1992, for example, curbside pickup programs expanded almost sevenfold across the United States. The public financing of this infrastructure acted as a corporate subsidy: recycling allowed the beverage and packaging industries not to bear the cost of the waste they produced. Government funds, together with individuals performing the therapeutic ritual of recycling, subsidized the cost of industry waste. Presented as a panacea to the nation's waste crisis, recycling marginalized other proposed solutions—bottle bills, bans on nonreturnable packaging, and policies that would require polluters and waste producers to pay for environmental cleanup operations. Recycling offered publicly financed feelings of environmental hope and provided consumers with affective empowerment, but also reinforced unsustainable industry practices. As an emotionally rewarding therapeutic ritual, recycling provided a template for neoliberal environmental citizenship and seemed to offer a resounding answer to the question posed by the Earth Day 1990 slogan: "Who says you can't change the world?"[17]

Yet, even as many mainstream environmental groups endorsed the Earth Day message, other activists challenged the relentless focus on green consumerism and individual action. "By focusing attention on the smaller issues in our personal lives," one critic argued, "we get to feel good about ourselves and close our eyes to the larger structural issues which we sometimes feel powerless to change." Likewise, Barry Commoner worried that the emphasis on personal action "confuses feeling good with having an influence on the problem." According to these critics, Earth Day 1990 reduced environmentalism to a popular form of therapy rather than a political struggle to create long-term sustainability. Recycling helped alleviate environmental anxiety and made individuals feel like virtuous consumers, yet this brand of environmental hope failed to address the structures of power that caused the environmental crisis.[18]

Commoner and other critics looked to the burgeoning environmental justice movement as a more promising model for environmental politics. Two decades earlier, during the period surrounding Earth Day 1970, subaltern movements had challenged mainstream framings of environmentalism by emphasizing the power relations that determined the unequal levels of ecological risk. Rather than accepting the popular discourse of universal vulnerability, these activists demonstrated how different groups of Americans, especially the poor and racialized minorities, experienced conditions of ecological injustice and faced problems—including lead poisoning in the inner cities and farmworker exposure to pesticides—that were often obscured by media portrayals of the environmental crisis. Grassroots environmental activism, especially among people of color, expanded dramatically during the 1980s and led to sharp criticisms of the mainstream environmental movement and US environmental policies. In 1987 the Reverend Benjamin Chavis, a longtime civil rights activist, coined the term "environmental racism" to name a set of interlinked practices that produced ecological inequities. According to Chavis, the term encompassed "racial discrimination in environmental policy-making and the enforcement of regulations and laws, the deliberate targeting of people of color communities for toxic waste facilities, [and] the official sanctioning of the life-threatening presence of poisons and pollutants in our communities."[19]

The organizers of Earth Day 1990 hoped that the twentieth-anniversary celebration would help mainstream environmentalism shed its predominantly white image. To do so, they made a "special effort to involve U.S. minorities" in various events leading up to Earth Day itself. In particular, Denis Hayes accompanied the Reverend Jesse Jackson, the director of the National Rainbow Coalition and the most high-profile civil rights leader of the 1980s, on a toxic tour of several sites—including Richmond, California; Albuquerque, New Mexico; and Louisiana's "Cancer Alley"—where minority communities were organizing and struggling against the poisoning of their local environments. At one rally, Jackson placed environmental justice within the longer history of civil rights activism, including struggles to end segregation and to improve housing conditions for inner-city residents. "None of those rights is more basic than the right to breathe free," he asserted. "For unless I have the right to breathe free, the right to drink good drinking water, no other right can be realized." Despite Jackson's media prominence, this Earth Day–sponsored toxic tour attracted almost no attention from national news outlets. While local newspapers covered their various stops, not a single segment of ABC, CBS, or NBC News reported on the tour or grappled with its effort to frame environmental justice as a citizenship right.[20]

Even as grassroots activists protested environmental racism, even as the United Church of Christ Commission for Racial Justice released a major report documenting that people of color were disproportionately exposed to toxicity, and even as antitoxic groups challenged mainstream environmental organizations for ignoring social justice issues, the news media, for the most part, packaged Earth Day as a generic effort to mobilize environmental feeling. Popular environmentalism appeared as an enormous exercise in affective empowerment rather than a quest for community empowerment, as a campaign for recycling and green consumerism rather than a challenge to corporate power structures.[21]

Just as Earth Day 1990 organizers sought to incorporate "a new rainbow of hues" into the environmental movement, *The Earth Day Special* featured a multicultural cast to represent the racial diversity of the citizenry. Unlike previous examples of popular environmental imagery that presented white people as symbols of universal vulnerability, *The Earth Day Special* offered a more variegated portrait of the nation. In the aftermath of civil rights struggles and the rise of multiculturalism, the national community could no longer be represented by white bodies alone. Yet though *The Earth Day Special* incorporated people of color into the environmental imaginary, their presence did not convey difference and division with the citizenry or reveal the power relations that determined the levels of environmental risk. Rather, their appearance expressed a unifying, race-neutral vision of the environmental crisis. In one scene, a large group of African American hip-hop artists performed a song titled "The Earth Raps Back." While the rappers themselves signified "blackness" and the "inner city" to mass audiences, their lyrics did not, in any way, indicate the disproportionate effects of pollution and toxicity on minority communities. Likewise, African-American cast members of *The Cosby Show*—the most popular sitcom of the 1980s, featuring the affluent Huxtable family—vowed to change their everyday behavior by taking shorter showers and leaving pot lids closed while cooking. In the multicultural environmentalism envisioned by *The Earth Day Special*, racially diverse citizens did not protest inequities, but rather demonstrated their ecological virtue through individual action.[22]

While environmental justice advocates spoke at some major Earth Day rallies, news media coverage emphasized the alluring presence of celebrities rather than the politics of grassroots activism. In New York's Central Park, for example, Peggy Shepard, who cofounded the environmental justice group West Harlem Environmental Action, "noted that pollution was an ally of racism." Though Shepard explained how "environmental contamination strikes hardest in black and Hispanic communities," that night all three networks instead showed footage of Hollywood heartthrob Tom Cruise, who, at a rally in Washington, DC, screamed into a microphone: "Have you planted a tree? Do it!"[23]

If Cruise and other celebrity environmentalists emphasized personal micro-solutions to the environmental crisis, other widely circulated images, including the official logo of Earth Day 1990, sought to represent the global macro-perspective (fig. 15.3). "Obviously, the symbol to use was the globe," logo designer Scott Mednick explained. "But whichever side of the globe you show, you're going to upset someone by not showing the other side." For Mednick, the iconic whole earth photograph seemed inadequate. To express a truly global vision, he wanted to place all nations in the same pictorial space, to squeeze them into a two-dimensional circle so that the spectator could apprehend them all simultaneously. "I call it my 'Last Supper' solution," he explained. "I could never understand why everyone in the Last Supper was eating on the same side of the table." The Earth Day logo flattened all differences, placed the entire world at the same table, and made it appear that national political divisions and global asymmetries of power no longer mattered. With the easing of Cold War tensions, many observers joined Mednick in hoping that people from a diverse array of countries would soon unite in a common project to protect the environment.[24]

The post–Cold War frame performed complex ideological work in this context. For many viewers, photographs of the fall of the Berlin Wall signified hope, a sense of expansive possibility about the capacity of people to change the world. Yet this uplifting coverage would

FIGURE 15.3. Earth Day 1990 logo by Scott Mednick. Courtesy of Scott Mednick.

also be joined by disturbing portraits of ecological devastation in the former Soviet bloc. In the time leading up to and following Earth Day, the news media focused public attention on environmental tragedies across the region. From Copsa Mica, Romania, dubbed by *Time* magazine "the blackest town in the world," a place where, CBS News explained, people constantly encountered "a level of pollution unthinkable in America," to the vast area contaminated by the 1986 Chernobyl nuclear power accident, these reports emphasized the horrors of Soviet-style development. The writer Garry Wills commented on the emotional power of the "gruesome photographs" *Time* featured in one of its pre–Earth Day stories about the radioactive legacies of Chernobyl. "Have you ever seen an eight-legged colt?" he continued. "The one pictured in *Time* looks like something out of a medieval tale of witchcraft. An eyeless pig is there as well. These are not fakes, or the odd freaks to be looked at in a sideshow; they are the things growing up all around people who know that the same infernal twistings are taking place in their own children."[25]

Likewise, a week after Earth Day, the *New York Times Magazine* featured a photo essay titled "Eastern Europe: The Polluted Lands." The cover photograph showed an elderly woman and a trio of children in Copsa Mica. Despite the apparently rural setting, soot blackened the sky and darkened their faces. Looming in the background, a coal-powered rubber factory spewed noxious pollutants into the atmosphere, endangering human health and the environment. This image resembled Martha Cooper's photograph of a mother and child near the cooling towers of Three Mile Island (see fig. 9.2), as both depicted bucolic landscapes and children's bodies imperiled by industrial technology. However, while the nuclear power plant posed an invisible threat to Pennsylvania residents, the black clouds surrounding Copsa Mica represented an extremely visible form of environmental danger.[26]

The US media repeatedly observed that for all the ecological sins of consumer capitalism, at least this system did not lead to Chernobyl or Copsa Mica. These reports vividly demonstrated the significance of democracy to environmental protection: denied a voice in policy decisions, citizens in Soviet-controlled areas suffered from the ruthless development strategies imposed by Communist central planners.

Yet, for all their documentary truth value and ability to reveal the ecological effects of Soviet domination, these images also operated in larger cultural and political fields that made the market and democracy seem inextricably linked. As market metaphors became increasingly dominant in American public life, consumer choice appeared as a crucial form of democratic expression. The fading of Cold War tensions led Francis Fukuyama to proclaim the "end of history," and to see in the triumph of liberal-democratic capitalism the hopeful glimmerings of a post-ideological global future.[27]

Popular discussions of the planetary environmental crisis, the end of the Cold War, and economic globalization all made the world seem increasingly interconnected. Yet global market forces—as exemplified by the rise of the hazardous waste trade—also enabled the United States and other rich countries to dump toxic waste onto the poorer nations of the global South, thus displacing environmental risk unevenly. As the historian Emily Brownell argues, "With the public's attention fixed on the issue of toxic waste in their backyards and neighborhoods, many industries started to take their waste overseas to countries . . . where disposal was unbelievably cheap and generally unregulated." At the global level, these practices grew out of neocolonial relationships and stark power imbalances between Northern and Southern countries. At the level of US environmental politics, the ongoing disposal of hazardous waste abroad revealed the failure to reduce toxics production. "Waste trading acted as a pressure-valve," Brownell concludes, "allowing the United States to avoid seriously addressing its waste problems."[28]

While Vic, played by Danny DeVito in *The Earth Day Special,* worried about the world his grandchildren would inherit, this focus on familial futurity ignored the geographical outsourcing of environmental risk. As in other expressions of popular environmentalism, DeVito evoked an image of innocent children to warn of long-term environmental danger. Yet "what the grandchild approach obscures," the literary critic Rob Nixon notes, "is the role of transnational, often imperial economic practices whereby someone else's grandchildren—in a distant land and a distant decade—will be inheriting the problems that affluent people in the here and now outsource to them." *The Earth Day Special*'s depiction of environmental time as a personal, familial concern thus failed to consider the global frameworks of power that determine the transnational inequities of danger. Despite the post–Cold War rhetoric of universality and one-world environmentalism, the accelerating effects of globalization and neoliberal economic practices have intensified and exacerbated the long-term disparities of environmental risk.[29]

Further, the "optimistic afterglow" of the post–Cold War moment did not translate into meaningful environmental change on the global level, especially in relation to the accretive dangers of greenhouse gas emissions. At the UN-sponsored "Earth Summit" held in Rio de Janeiro, Brazil, in 1992, US delegates sent by George H. W. Bush stridently insisted that "the American life-style is not up for negotiation." The global marketplace—together with the short-term focus on perpetual prosperity and faith in boundless economic growth—trumped environmentalist warnings about the hazards of climate change.[30]

Even as the media extravaganza of Earth Day 1990 and other popular framings of environmentalism acknowledged the long-term, accumulating dangers of toxicity and other environmental problems, the specific dangers of fossil fuel dependency faded from view. Just as Bill McKibben and Murray Bookchin had feared in the aftermath of the *Exxon Valdez* disaster, media temporality failed to grapple with the slow violence of climate change. As the environmental writer Mark Hertsgaard noted in 1989, global warming does not "yield daily photo opportunities tailored to the eye-blinking attention span of the media. How do you take a picture of the Earth getting hotter?" Even when the news media described various slow-motion disasters in the making, the delayed effects of global warming did not gain the same level of visibility accorded to other environmental dangers. While Earth Day and the Earth Summit represented major milestones in the history of global environmental concern, the economic imperatives of globalization, together with neoliberal deregulatory fervor, promoted short-term profit rather than long-term sustainability.[31]

Near the end of their *Earth Day Special* skit, Meryl Streep followed Kevin Costner's advice by tossing her beer can into the recycling bin. Costner then asked her, "Do you feel better?"

"I do," she said, and then added, with a flirtatious giggle, "But I'm not sure it has anything to do with the environmental movement."

Rather than continuing this banter, though, they concluded by solemnly recognizing their duty to act as proper environmental citizens. Throwing the aluminum can into the bin offered a tangible manifestation of virtuous consumption and provided a glimpse of ecological harmony generated by personal responsibility. While 1980s culture had celebrated greed and conspicuous consumption, *The Earth Day Special* joined other popular media texts in urging Americans to participate in a movement to minimize the impact of consumer culture and envision ecological sustainability.

Rather than representing a sharp break from Reaganism, Earth Day 1990 instead smoothed the transition to a green vision of neoliberal citizenship: one in which the market—and not the government—became the prime mechanism for guaranteeing the rights of citizens to a clean environment. Indeed, the Clean Air Act of 1990—the main legislative victory that accompanied Earth Day—typified market-oriented environmental policy. While grassroots activists in the antitoxics movement challenged the prerogative of industry to poison the environment, and called for sharp reductions in the production of toxicity, the Clean Air Act, supported by several leading mainstream environmental groups, created an emissions trading system that allowed corporations to buy and sell pollution as a commodity.[32]

The Clean Air Act also failed to regulate greenhouse gas emissions or to mandate higher gas mileage for motor vehicles. Indeed, the average fuel economy of passenger vehicles in the United States declined throughout the 1990s, while the nation's dependence upon imported oil continued to rise. In August 1990, less than four months after Earth Day, Saddam Hussein sent Iraqi troops to invade Kuwait. This military action, the political scientist Michael Klare explains, would immediately be viewed by Bush and his advisers "through the lens of the Carter Doctrine: as a threat to Saudi Arabia and the free flow of oil from the Gulf." The Gulf War of 1991 fulfilled the principles of the Carter Doctrine and demonstrated that the United States would use "any means necessary, including military force" to safeguard its access to Persian Gulf oil.[33]

While Denis Hayes and other Earth Day organizers hoped that the event would launch the "green decade," the 1990s instead became the decade of the sport utility vehicle (SUV). Sales of these gas-guzzling vehicles soared throughout the decade as the federal government—under the presidencies of both Bush and Clinton—chose not to require SUVs to meet higher fuel efficiency standards. Greenhouse gas emissions continued to stream into the atmosphere in ever-growing quantities. Far from becoming an international environmental leader, the United States instead remained the world's greatest contributor to global warming.[34]

The American failure to address the impending climate crisis in a more responsible fashion stems from numerous factors, including the right-wing, corporate-sponsored effort to foster uncertainty by denying the scientific consensus on the anthropogenic causes of global warming. In addition, the popular framing of environmentalism as a market-oriented, green-consumerist strategy also contributed to the national neglect of climate change and other accretive disasters in the making. Indeed, the limits to environmental reform were embedded within these depictions of the movement as a personal, consumerist response to crisis. While the repeated emphasis on individual responsibility furthered the popular commitment to recycling, while pesticide fears boosted sales of organic food and led Whole Foods and other high-end grocers to expand their market share, the neoliberal template of environmental citizenship did not call upon the state to limit greenhouse gas emissions or curb the power of globalizing corporations. Americans instead were urged to recycle and, if they could afford it, to shop their way to ecological salvation. In a period marked by rising rates of economic inequality, green consumerism catered to the affluent and obscured questions of power relations. The structural, systemic assault on the ecosphere continued; the release of greenhouse gases and the production of toxicity escalated; and the poor and racial minorities within the United States, together with people in the global South, experienced higher levels of ecological risk. All of this happened while recycling programs expanded across the United States, while green consumer products promised to shield the affluent from harm, and while recycling and other individual actions provided Americans with a therapeutic dose of environmental hope.[35]

The Strange Career of
An Inconvenient Truth

"Former Vice President Al Gore is starring in a new documentary about global warming," Jay Leno informed *Tonight Show* viewers. "I believe it's called," he continued, but then pretended to doze off—snoring loudly as the studio audience guffawed. The prospect of the former politician, frequently derided as "stiff" and "wooden" by the press, lecturing moviegoers about climate change no doubt seemed potentially soporific. After seeing *An Inconvenient Truth* (2006), numerous reviewers therefore felt compelled to reveal their surprise at just how entertaining and enthralling they found the film—and Al Gore—to be. "Boring Al Gore has made a movie," one critic began. "It is on the most boring of all subjects—global warming. It is more than 80 minutes long, and the first two or three go by slowly enough that you can notice that Gore has gained weight and that his speech still seems oddly out of sync. But a moment later, I promise, you will be captivated and then riveted and then scared out of your wits." Another critic noted, "*An Inconvenient Truth* is full of surprises. . . . The film offers viewers an emotionally rich, visually entertaining story. Also a little shocking is that Al Gore is such a compelling central character."[1]

The movie's critical acclaim, including its Academy Award for Best Documentary Feature, contributed to Gore's phoenix-like rebirth in American (and global) public culture. Some even described Gore as a "rock star" who had transitioned "from failed presidential contender . . . to the most unlikely of global celebrities." "He's an icon," a leading Republican strategist observed. "Imagine that: Al Gore, Mr. Straight and Narrow, Mr. Dull on Wheels—now he's culturally *cool*." Gore soon became the world's most visible carbon warrior. He went on tour, giving his slide show lecture to packed houses in the United States and around the world. He appeared on the cover of numerous magazines, earned cameo roles on popular TV shows, and stood onstage with Leonardo DiCaprio to announce that the Oscars had gone green. Gore's environmental fame also enabled him to become a co-recipient of the 2007 Nobel Peace Prize, awarded jointly to the Intergovernmental Panel on Climate Change, for documenting and popularizing knowledge of anthropogenic climate change. Gore thus became the only person ever to win a Nobel Peace Prize based largely upon his role in a movie.[2]

The surprising popularity of *An Inconvenient Truth* marked a watershed moment in public understanding of climate change. In previous chapters we have seen how the systemic dangers of fossil-fuel dependency—especially the accretive catastrophe of climate change—have remained invisible in media coverage that tends to emphasize the sudden violence of oil spills or the shocking onset of fuel shortages. *An Inconvenient Truth* challenged the conventions of media temporality by visualizing the long-term, escalating dangers of global warming. More than any other cultural text, this film allowed spectators to see global warming as an imminent reality, and to feel viscerally connected to an issue that previously seemed too far removed—in both space and time—from their everyday experience of the environment.

While many journalists, film critics, and others have marveled at the strange career of *An Inconvenient Truth*, this discussion has failed to consider the film's place in the longer history of environmental icons. Indeed, much that seemed novel about the film drew upon tropes and representational strategies that have repeatedly popularized and delimited the scope of American environmentalism. The film's fusion of fact and feeling, its framing of universal vulnerability and responsibility, and its failure to address power relations and environmental inequities: we have seen similar achievements and blind spots in other images. From this vantage point, *An Inconvenient Truth* can be considered both surprisingly innovative and disappointingly familiar: a popular text that visualized the climate crisis but also reproduced the problems and limits of previous environmental icons.

Throughout *Seeing Green*, I have traced the emotional history of popular environmentalism by showing how a series of images—from the

Dr. Spock ad to the Alar campaign—combined scientific knowledge with feelings of fear to warn about threats to the environmental future. Similarly, *An Inconvenient Truth* sought to mobilize audience concern through the emotive presentation of scientific data. In the film, Gore brought personal and familial feelings into public life, and merged affect with climatology to picture global warming's threat to futurity.

Until 2006, media coverage had tended to distance and displace global warming, made it seem abstract, a theoretical forecast rather than a concern for the foreseeable future. That changed just before the release of *An Inconvenient Truth*. In the months leading up to the film's debut, emotionally saturated media coverage brought increasing visibility to the climate crisis, and global warming acquired its first iconic image. A brief review of these images will allow us to appreciate how *An Inconvenient Truth* expanded the visual repertoire of climate change and joined other environmental icons in conveying scientifically informed fears of an escalating crisis.

"Be Worried. Be *Very* Worried." So warned *Time* magazine in an April 2006 cover story on global warming. The cover featured a photograph of a lone polar bear perched on floating ice, gazing uncertainly at the surrounding sea (fig. 16.1). The polar bear soon became the most recognizable image of climate change, circulating in many media forms. That summer, *An Inconvenient Truth* included a cartoon sequence in which a polar bear tries, repeatedly and unsuccessfully, to climb onto chunks of melting ice (fig. 16.2). The animated creature seems destined to drown. The next spring, for the cover of *Vanity Fair*'s second annual "Green Issue," the photographer Annie Leibovitz depicted Hollywood heartthrob Leonardo DiCaprio standing on an Icelandic glacier, along with a digitally added image of Knut, a popular polar bear cub from the Berlin Zoo.[3]

Polar bear imagery transcended the ephemeral conditions of weather to reveal alarming evidence of climate change, to make this long-term, gradually escalating problem appear immediate, located in the coeval present rather than in some vague, imaginary future. The lone polar bear hewed closely to the visual tradition of depicting individual animals as archetypal victims of fossil fuels—think of the oil-stained infant sea lion near Santa Barbara in 1969 or the sea otter lying dead along the Alaskan coast in 1989. Unlike oil spill victims, though, whose tragic fate could be read as merely accidental, the polar bear's plight required the media to move beyond dramatic moments of disaster to consider instead the temporality of climate change—the product of accretion rather than accident. To present the polar bear as poster

FIGURE 16.1. *Time* magazine cover, April 3, 2006. Copyright 2006 Time Inc. Used under license. *Time* and Time Inc. are not affiliated with, and do not endorse products or services of, licensee.

child of global warming, the media emphasized how the creature's surrounding ecosystem began to succumb to the damaging effects of climate change. "A polar bear negotiates what was once solid ice," *Time*'s caption explained. "Bears are drowning as warmer waters widen the distance from floe to floe."[4]

FIGURE 16.2. Animated polar bear and melting ice. Frame capture from *An Inconvenient Truth*, directed by Davis Guggenheim, 2006.

As the polar bear emerged as global warming icon, the Ad Council, in conjunction with the Environmental Defense Fund (EDF), launched a campaign that depicted the vulnerable child as another innocent victim of climate change. In one spot a middle-aged white man stands in the path of an onrushing train. "Global warming," he says. "Some say irreversible consequences are thirty years away." As the train speeds toward him, the man calmly continues, "Thirty years? That won't affect me." He steps off the track—only to reveal a young white girl about to be struck by the barrelling locomotive. Unaware of the looming catastrophe, the girl, with blond curls and a white dress, stands motionless on the track. Innocent and completely devoid of agency, she cannot protect herself from the climate crisis. The Ad Council and EDF reference a familiar trope in environmental imagery to picture the child as an emotional emblem of the future. Here, though, the next- generation discourse refers not to a specific hazard—not to the risk of strontium 90 or the menace of air pollution, not to the fear of the China syndrome or the danger of Alar on apples—but rather to a much broader threat to the Earth's climate systems triggered by the ongoing accumulation of greenhouse gases in the atmosphere. By placing the child in front of the speeding train, the ad spectacularizes the unspectacular: it makes the accretion of greenhouse gases appear as a sudden, visible threat and establishes a temporal and emotive bond between present inattention and future danger.[5]

Likewise, *An Inconvenient Truth* offered a filmic fusion of Gore's slide show lecture with his life story to personalize and emotionalize the climate crisis. Indeed, before viewers see Al Gore or his slide show, before they encounter disturbing examples of planetary danger, the film's first visuals reveal bright green leaves, dappled by the sun, growing on a dense cluster of trees alongside a tranquil river (fig. 16.3). "You look at that river, gently flowing by," Gore observes. "You notice the leaves, rustling with the wind. . . . It's quiet. It's peaceful. And, all of a sudden, it's a gear shift inside you. And it's like taking a deep breath and going, 'Oh yeah, I forgot about this.'"

In this opening sequence, the pictured place goes unnamed, yet viewers may rightfully suspect that the location holds personal significance for the narrator—and, as they find out later in the film, it turns out to be a riverbank on the family farm near Carthage, Tennessee, where Gore spent much of his childhood. As a recurring motif in the film, the meandering river personalizes Gore's connection to nature. In another sequence, the director Davis Guggenheim intersperses black-and-white family photographs—including pictures of Gore as a young boy—with contemporary footage of the river and farm shot in Kodachrome film. The bright, saturated colors of Kodachrome, together with Guggenheim's intentionally jerky camera style, lend an amateurish, home-movie feel to the footage. The images appear as authentic family mementoes, nostalgic representations of Gore's past reveries in

FIGURE 16.3. Caney Fork River, Tennessee. Frame capture from *An Inconvenient Truth,* directed by Davis Guggenheim, 2006.

this environment. "He talks about lying beside a river on a lazy summer day—the commonplace idyll of a lone person in a tranquil ecstasy, utterly at home in nature," one reviewer observed. "[The film] begins that way, and each time Gore returns to this enraptured mood, after a procession of nightmares and dangers, it has greater resonance."[6]

The personal and the planetary merge in an autobiographical sequence, when Gore recounts the car accident that nearly killed his six-year-old son. Viewers see photographs of family members visiting the young boy in the hospital, as Gore comments in a voice-over: "The possibility of losing what was most precious to me. I gained an ability that maybe I didn't have before." As he continues to talk, images of the river—shrouded in mist and floating lazily along—return to the screen. "But, when I felt it," Gore says, "I felt that we could really lose it. That what we take for granted might not be here for our children." The frightening accident and his son's painful recovery; the perils of global warming and the potential threat to the river; Gore's renewed concern—as a parent and a politician—for climate change: all these images and personal experiences fuse in *An Inconvenient Truth* to picture children and nature as emotional signs of the future and as the prime victims of the catastrophic effects prophesied by climate scientists. While some critics worried that the treatment of "family tragedies" might "strike some viewers as maudlin notes from a campaign biography," most found Gore's focus on futurity and his effort to fuse "the personal and the political" to be "moving" and "powerful."[7]

Near the end of the film, *An Inconvenient Truth* returns, for a final time, to the river. As the sun casts a ravishing light across the water, Gore delivers his closing words: "Future generations may well have occasion to ask themselves, 'What were our parents thinking? Why didn't they wake up when they had a chance?' We have to hear that question from them—now." The river provides a personal site in which Gore can muse about the intergenerational imperative of environmental action, a theme noted by many film reviewers. "In 39 years, I have never written these words in a movie review," the legendary film critic Roger Ebert proclaimed, "but here they are: You owe it to yourself to see this film. If you do not, and you have grandchildren, you should explain to them why you decided not to." Like previous environmental icons, *An Inconvenient Truth* frames environmental citizenship as an emotive project that takes the future into account.[8]

In addition to merging personal and planetary concerns, *An Inconvenient Truth* develops other innovative strategies to visualize the causes of global warming. Placing climate history in an extensive time frame, stretching back hundreds of thousands of years, the film does not re-sort to textbook recitation of data but rather infuses scientific fact with feeling to engender an emotional response in audiences. In some of the film's most striking and frequently commented upon moments, Gore and Guggenheim use graphs to display quantitative information in a surprisingly captivating fashion. Statistics do not appear as lifeless numbers but rather act as fear-inducing signs of environmental catas-trophe, scientific information that simultaneously addresses viewers as emotive and cognitive beings. *An Inconvenient Truth* stages scientific data as a narrative of human-induced environmental harm, a visual story in which two sets of quantitative information—carbon dioxide levels and average global temperatures—generate shock and fear in the capacity of industrial society to transform the climate.

On a large horizontal graph, a jagged red line tracks the changing levels of carbon dioxide in the atmosphere over the past 650,000 years. Throughout this vast expanse of time the carbon dioxide line fluctu-ates up and down, between two and three hundred parts per million. Then a jagged blue line plots the average temperature readings over the same period. Viewers immediately notice the uncanny similarity be-tween the two lines: every rise in the red line seems to be matched by a corresponding rise in the blue line, and every dip in the red generates a corresponding dip in the blue. Gore's visual rendering of scientific information evoked emotional responses from audiences. "I can't think of another movie in which the display of a graph elicited gasps of hor-ror," the *New York Times* film critic A. O. Scott observed, "but when the red lines showing the increasing rates of carbon-dioxide emissions and the corresponding rise in temperatures come on screen, the effect is jolting and chilling."[9]

To heighten the emotional impact of the graph, Gore emphasizes the humanistic meanings of global warming. He stands on a mechani-cal cherry picker that elevates him to the current carbon dioxide level (fig. 16.4). Before the carbon dioxide amounts have bounced up and down, but now the red line moves relentlessly upward. "In the next fifty years," Gore observes, "it's going to continue to go up. . . . When some of these children who are here are my age, here's where it's going to be." By referring to children, Gore presents the statistics as menac-ing markers of ecological risk that pose a grave threat to futurity. His ride on the cherry picker, according to one of the film's producers, con-stitutes *An Inconvenient Truth*'s "most terrifying moment. Look at all

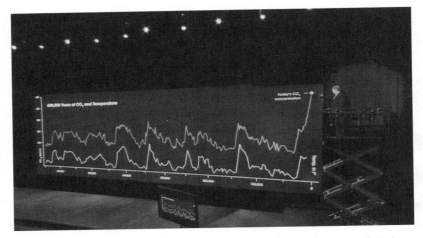

FIGURE 16.4. Al Gore and graph with CO_2 and temperature data. Frame capture from *An Inconvenient Truth*, directed by Davis Guggenheim, 2006.

this history and then at the near-term future, at the changes happening now. If there's one visual that audiences can take away from this movie, it's this one."[10]

Although Gore joins *Time* and other mainstream media in evoking pity for the ice-deprived polar bear, he focuses more attention on the plight of people. He recounts recent weather events—from heat waves in Europe to typhoons in Japan—to demonstrate how increasing temperatures in the atmosphere and oceans have produced cataclysmic harm. In another viewing context, such as news media coverage that lacks Gore's historical perspective, this imagery of storms, floods, and droughts might be seen as freakish events of nature, devoid of any larger meaning, unrelated to accumulating carbon emissions. Here, though, the pictures appear as the calamitous result of the graph's ascending red line and resemble, as Gore puts it, "a nature walk through the book of Revelations."

Following these photographs of recent disasters, Gore shifts to the future by showing maps and animated images that warn of dramatic sea level rise. The rising waters redraw the maps of the world—from the Netherlands to India—and flood large parts of Florida and the San Francisco Bay area. Gore pauses to linger over his last example: a map of lower Manhattan. "After the horrible events of 9/11, we said never again," he observes. "But this is what would happen to Manhattan." Water begins to flow through the streets of the animated map, charting new pathways of destruction (fig. 16.5). "The area where the World Trade Center Memorial is to be located would be under water," Gore concludes. "Is it possible that we should prepare against other threats besides terrorists? Maybe we should be concerned about other problems as well." Gore evokes the tragedy of 9/11 but also encourages audiences to look beyond the spectacular violence of burning towers. Just as previous environmental icons merged facts with feelings to warn of long-term, incremental crises, Gore's display of emotion-eliciting graphs, photographs, and other images works to extend temporal horizons into the future and to invite viewers to foresee the prospective calamities of global warming's slow violence.[11]

Throughout *Seeing Green*, I have argued that environmental icons create a picture of universal vulnerability to enframe all people as equally susceptible to ecological harm. From SANE ads to gas mask imagery, from the cooling towers at Three Mile Island to the fear of Alar on apples, this trope has popularized environmental anxiety but also masked the structural inequities that produce ecological injustice. Similarly, *An Inconvenient Truth* depicts universal vulnerability and draws on previous environmental icons, including the *Earthrise* photograph and other earth imagery, to emphasize the planetary dimensions of the climate crisis (fig. 16.6). The whole earth, the stranded polar bear, and the innocent child all provide Gore with resonant images of universal

FIGURE 16.5. Animated map of lower Manhattan. Frame capture from *An Inconvenient Truth*, directed by Davis Guggenheim, 2006.

FIGURE 16.6. Al Gore and earth image. Frame capture from *An Inconvenient Truth*, directed by Davis Guggenheim, 2006.

vulnerability and convey a generalized sense of danger. This planetary perspective, though, conceals the specificity of risk and ignores the local particularities of place and power that render some people more vulnerable than others to environmental hazards. Indeed, while Gore wants to foreground the humanistic dimensions of global warming, he repeatedly effaces the structural inequities and local ecologies that shape the experience of risk. Gore's fixation on the universalizing image of the whole earth blinds him to the markedly differentiated experiences on the ground, to the ways in which global warming poses unequal levels of danger to different human communities.[12]

Like other mainstream sources that mobilize the polar bear as an iconic victim of climate change, *An Inconvenient Truth* completely ignores the plight of Arctic indigenous peoples whose cultures and landscapes are facing profound changes produced by melting polar ice. Even as Gore and Guggenheim summon the pathos of viewers by picturing the risk to charismatic megafauna, the film neglects the human costs of global warming in the region and fails to mention the presence of Arctic communities and their ongoing efforts to struggle for climate justice. On Earth Day 2005, for example, just over a year before the release of Gore's film, Inuit activists "used their bodies to trace the outline of an Inuit figure on the ice sheets of Baffin Island, a figure that was accompanied by the words 'Arctic Warning.'" This event, together with other forms of indigenous activism, have emphasized how escalating levels of carbon emissions, largely produced by areas far from the

Arctic, are exerting strikingly visible effects—the dramatic melting of sea ice and permafrost, the disruption of caribou migration patterns, and the flooding of coastal villages—on the region's ecological systems and human communities.[13]

Gore's treatment of Hurricane Katrina reveals a similar blind spot. He describes the hurricane as, in part, a human-induced catastrophe, its extreme violence fueled by the warming waters of the Gulf of Mexico. He places the increasing intensity and frequency of hurricanes within the longer temporal frame of rising carbon emissions and considers Katrina a harbinger of more destructive storms to come. The hurricane also figured prominently in the film's promotional imagery: *An Inconvenient Truth*'s most widely distributed poster featured an image of swirling plumes emanating from smokestacks and taking recognizable form as the ominous spiral of a hurricane. The poster visually reinforces the message of the movie: more emissions lead to more hurricanes.

Still in production when Katrina made landfall, *An Inconvenient Truth* incorporated many of the disaster's shocking media images. While Gore places the catastrophe within the longer timescale of carbon-induced climate change, he ignores the local histories of power and geography that determined how different groups experienced the calamity. For all of Gore's efforts to historicize environmental crisis and to view Katrina and other calamities within a longer narrative of rising carbon emissions, his framing of the hurricane's impact on New Orleans nevertheless replicates the foreshortening of time produced by mainstream media accounts of natural disaster. While Gore questions the "naturalness" of Katrina by linking it to anthropogenic climate change, he fails to consider the city's decades-in-the-making structural inequities that shaped the uneven levels of vulnerability to the storm. The concentration of minorities in the Ninth Ward and other low-lying areas, the evisceration of public services in the neoliberal era, and the neglect of the city's infrastructure all conspired to turn Katrina's landfall into an enormous tragedy of environmental injustice.[14]

Although Gore seeks to move beyond shortsighted media coverage that ignores deeper patterns of environmental change, he focuses on human-nature relationships at the macro-level, on the large-scale forces that are transforming global climate systems, but neglects the historical processes that generate severe inequities in the distribution of environmental risk. For Gore, both melting ice in the Arctic and Katrina's devastation of New Orleans signify a universal story of environmental decline and demonstrate accelerating vulnerability in a warming world. In both cases he appeals to audiences as homogeneous members of the global community, bounded together by their planetary consciousness. Like other environmental icons that reference the trope of universal vulnerability, *An Inconvenient Truth* ignores the systemic practices and spatial inequities that have shaped—and will continue to shape—the markedly uneven human experience of global warming.

An Inconvenient Truth draws on a final trope examined in *Seeing Green*—the discourse of individual responsibility—to equate green consumerism with environmental citizenship. Following the pattern established by Pogo, the Crying Indian, and the "Don't Be Fuelish" campaign, Gore prescribes individual actions to ward off the looming crisis. Much like the Ad Council's recycling campaign, Kevin Costner's advice to Meryl Streep in *The Earth Day Special*, and other popular images produced around Earth Day 1990, he presents individual consumers as empowered players in the market, ensuring sustainability through their purchasing decisions. Through his focus on virtuous consumption, Gore subscribes to neoliberal visions of environmental hope.

Yet many viewers, including some leading environmentalists, felt disappointed by the film's emphasis upon individual action. Indeed, while global warming deniers hurled all kinds of insults at Gore, some of the most searching criticisms of *An Inconvenient Truth* came from those who agreed in large measure with the former vice president's presentation of the crisis but lamented his failure to fashion a compelling, inspiring vision of the future. "I don't know about you," the popular food writer Michael Pollan observed, "but for me the most upsetting moment in *An Inconvenient Truth* came long after Al Gore scared the hell out of me, constructing an utterly convincing case that the very survival of life on earth as we know it is threatened by climate change. No, the really dark moment came during the closing credits, when we are asked to . . . change our light bulbs. That's when it got really depressing." According to Pollan, the "immense disproportion between the magnitude of the problem" and the "puniness" of Gore's proposed solutions—using energy-efficient light bulbs, carrying reusable bags to grocery stores, and, if we can afford to, buying hybrid vehicles—"was enough to sink your heart." These gestures, while useful in reducing our personal carbon emissions, ultimately seem tokenistic, hardly the stuff to inspire the remaking of entire nations and economies.[15]

The trouble with Al Gore, though, lies not simply in *An Inconvenient*

Truth's failure to envision large-scale environmental changes. Beyond the filmic text, Gore's brand of environmentalism has closely followed the neoliberal paradigm. In particular, his promotion of carbon offsets as a solution to the climate crisis privileges individual consumer action as the basis for environmental citizenship.

Still flush with Oscar success, Gore found that his green consumerism took an unexpectedly personal turn when a right-wing think tank exposed his suburban Nashville's home extravagant energy usage. The 2007 report charged that the carbon warrior consumed "more than 20 times the national average" of electricity, with annual utility bills that totalled almost thirty thousand dollars. A spokesperson for Gore immediately issued a response. "The bottom line is that every family has a different carbon footprint," she claimed. "And what Vice President Gore has asked is for families to calculate that footprint and take steps to reduce and offset it." The Gores, she continued, "use compact fluorescent bulbs and other energy efficient measures and then they purchase offsets for their carbon emissions to bring their carbon footprint down to zero."[16]

Gore's promotion of carbon offsets reflected his faith in green consumerism and market-oriented environmentalism. Rather than reducing emissions at the source, offsetting allows purchasers to fund carbon-mitigating activities—such as tree-planting and renewable energy projects—that promise to atone for their personal greenhouse-gas sins. Carbon offsetting relies on the trope of ecological interconnection to imagine, for example, a mango forest planted in India absorbing, almost immediately, the carbon dioxide produced elsewhere. The macroperspective afforded by planetary consciousness, though, obscures certain material realities—such as, in the case of forest projects, how long it takes trees to grow before they remove the targeted amounts of CO_2 from the atmosphere—and it does nothing to lessen actual greenhouse gas emissions.[17]

Unlike Jimmy Carter, whose 1979 "Crisis of Confidence" speech called for personal sacrifice to diminish energy consumption, Gore offers a consumerist fantasy of green salvation through carbon offsetting. Although he joins Carter in emphasizing individual action, Gore does not appear as the cardigan-wearing scold beseeching Americans to forgo their narcissistic consumer habits. His vision of carbon neutrality does not demand sacrifice, but rather offers the allure of a short-term solution to reduce one's carbon footprint. As the environmental writer Heather Rogers notes, the carbon offset represents a curious form of consumption—one that seems akin to online shopping but that also projects a reassuring, almost anticonsumerist aesthetic. "The website is the primary tool for carbon-offset firms," Rogers observes. "Here they promote themselves and sell their wares. . . . The sites telegraph responsibility and simplicity, the antithesis of consumption. The aesthetic is usually spare, lots of white space and green lettering, with photos of towering windmills and resplendent banks of solar panels." Just as the recycling logo connotes environmental sustainability, alternative energy icons such as windmills and solar panels now provide symbols of hope for consumers worried about carbon emissions. In contemporary cyberspace, ecological redemption is just a mouse-click away.[18]

The time frames of hope generated by the carbon offset contradict the long-term perspective that Gore emphasizes in his depiction of the causes of global warming. While Gore places anthropogenic climate change within an extensive temporal vision, his endorsement of carbon offsetting instead promises instantaneous environmental improvement. As the political scientist Kate Ervine notes, "The offset offers the perfect short-term solution by providing consumers with an immediate tool through which to address the problem." Yet the quick fix of green consumerism ignores the long-term problem of escalating emissions in the United States and other affluent nations through the process of geographic outsourcing, of paying "distant others, frequently located in the global south, to engage in emission reduction activities." Ervine explains: "Carbon offsetting ultimately represents the spatial displacement of responsibility for global warming mitigation from north to south." Gore's whole earth vision again ignores global divisions and power relations to enlist the undifferentiated, mythical "we" in a campaign to save the planet through virtuous consumption. His solutions merge the personal with the planetary, but obscure the geographies of risk and responsibility and enshrine faith in the market as the immediate savior, the short-term solution to the long-term accumulative crisis of climate change.[19]

The green consumerist nostrums that shopping will save the planet also became central to a slew of corporate advertising campaigns. In 2007, Philips Electronics joined Al Gore in presenting the compact fluorescent light bulb as the paragon of green purchasing. The TV commercial, developed by DDB—the successor agency to Doyle Dane and Bernbach, which famously created both "Dr. Spock is Worried" and "Daisy Girl" during the 1960s—deploys similar representational strategies to frame global warming as a threat to futurity. The commercial begins with images of melting glaciers as the female narrator links the scene to human-induced climate change. "The Arctic ice is shrinking at a record pace," the narrator explains. "Electricity used by lighting is one of the biggest sources of CO_2 emissions." Rather than glimpsing a lone polar bear on melting ice, audiences instead see a human baby, alone and adrift on an ice floe. Much as in "Daisy Girl," the camera then zooms in on the child's eye. As the baby looks up, his iris does not fill with the apocalyptic mushroom cloud, but instead reflects the sun's intensely glowing orb, which then, in the next shot, turns into an electric light shining on a city street, presumably a compact fluorescent manufactured by Philips. "New Philips energy efficient lighting uses less electricity than ever before, helping reduce CO_2 emissions," the narrator informs us. "Simplicity is a light bulb that can help change the world." The baby, no longer endangered in the Arctic, appears comfortably protected by his white mother and father, who smile and snuggle the child as they happily stroll together, the affluent nuclear family sheltered from environmental crisis by the panacea of green consumption. In the neoliberal vision promoted by Philips and Gore, environmental hope appears as the convenient truth of consumer culture.[20]

Thus far, I have stressed the limits to Gore's political vision, but the significance of *An Inconvenient Truth* cannot be reduced to the cinematic text alone. The cultural life of the film extends well beyond its images and narration—beyond what audiences see on movie, TV, or computer screens—to encompass its ineffable, unquantifiable, but no less real impact on environmental action today. We have seen how *The China Syndrome* (1979), despite its failure to portray antinuclear activism in a positive manner, became a surprising tool of protest. *An Inconvenient Truth*, despite its failure to envision large-scale solutions or promote action other than green consumerism, has brought much-needed legitimacy to various forms of climate activism. Indeed, without the popular success of *An Inconvenient Truth*, it would be difficult to imagine the rising public concern over global warming, the flourishing of groups such as 350.org (led by the writer Bill McKibben), and the massive acts of civil disobedience in opposition to the proposed Keystone XL Pipeline.

Through its emotive presentation of scientific data, *An Inconvenient Truth* rendered visible the seemingly invisible phenomenon of global warming. Challenging the conventions of media temporality, the film serialized and dramatized the extensive timescale of environmental crisis. In proposing solutions, Gore confined environmentalism to an individualist frame, a neoliberal quest for market-driven sustainability. Still, we should not overlook what Gore accomplished: More than any other cultural text, *An Inconvenient Truth* helped mass audiences glimpse the accretive crisis of climate change.

The burgeoning climate movement builds on Gore's success but seeks to move beyond his limited model of citizenship. Indeed, McKibben and other leaders of 350.org describe the group's founding in 2007 as a response to the "individual-driven message in vogue among environmentalists" following the release of *An Inconvenient Truth*. (The group takes its name from the amount of carbon dioxide—350 parts per million—that the climate scientist James Hansen claims is the "safe upper limit" for the atmosphere.) Rather than urging Americans to shop their way to ecological salvation, 350.org emphasizes that "'it's not light bulbs, not Priuses,'" but "'large systemic change' that is truly necessary" to reduce greenhouse gas emissions.[21]

For 350.org, images provide a tool to galvanize climate activism by helping people see themselves not as isolated individuals but as participants in a collective project. "We thought that one of the reasons why there was so little action against climate change was not because people weren't scared of it, but because it felt so big and each of us felt so small that it seemed pointless to do anything," McKibben explained. "So we decided that we would try to make people see that they were part of something very large." Here McKibben echoes Streep's conversation with Costner in *The Earth Day Special*—but transcends the therapeutic frame to promote collective political action. Hoping to foster a sense of global solidarity, 350.org leaders often encourage protest participants to upload their photos on Flickr and thereby see themselves as part of a larger struggle. After one major series of rallies, approximately twenty thousand photos arrived from around the world, including many that showed demonstrators standing en masse to form the number 350.[22]

In a video titled "Because the World Needs to Know" (2008), 350.org both borrows and departs from the visual politics of *An Inconvenient Truth*. The ninety-second animated video begins with carbon-emitting factories, cars, and chopped-down trees. Much like Gore's emotive display of scientific information, it then features a graph with a rapidly ascending red line: a warning that the earth's atmosphere has already exceeded the 350 threshold (fig. 16.7). Scenes of devastation follow, including rising flood waters, people wearing masks to survive the polluted atmosphere, and a blue planet turning fiery red. After about thirty seconds, though, the video shifts from eco-apocalypse to collective hope, showing people joining together to demand change. This

FIGURE 16.7. Frame capture from "350.org: Because the World Needs to Know." YouTube video by 350.org, 2008.

FIGURE 16.8. Frame capture from "The Day the World Came Together—The 350 Movement: October 24, 2009." YouTube video by 350.org, 2009.

video circulated widely—via the Internet and social media—in the time leading up to the group's first global day of action in 2009, in which 5,200 demonstrations took place in 181 countries (fig. 16.8).[23]

While solar panels and windmills figure prominently in 350.org iconography, contemporary climate activists do not present alternative energy as a depoliticized, techno-determinist solution. Unlike the Sun Day celebration in 1978, their actions—from protests against pipelines to fossil-fuel divestment—question the structure of dominant energy systems, especially the power of the fossil fuel industry. Unlike the Exxon boycott in 1989, they do not focus on one corporation as uniquely evil, but rather challenge an entire industry whose practices intensify the climate crisis.[24]

Harnessing a variety of new media, including Facebook, Twitter, and YouTube, Tar Sands Action and other groups have demanded that the Obama administration block plans to construct the Keystone XL pipeline that would convey oil distilled from Alberta tar sands to the United States. These activists have emphasized the urgency of climate change and organized a broad coalition of Americans—including students, rural landowners, civil rights activists, and indigenous groups—to oppose the pipeline. In late summer 2011, over the course of two weeks of sit-ins at the White House, 1,252 protesters were arrested; it was one of the largest acts of civil disobedience in recent US history. Al Gore, the centrist politician, the failed presidential candidate, the enthusiastic proponent of green capitalism, lent his support to the movement. Calling the tar sands "the dirtiest fuel on the planet," Gore urged Obama to cancel the Keystone project. "This pipeline would be an enormous mistake," he explained. "The answer to our climate, energy and economic challenges does not lie in burning more fossil fuels—instead, we must continue to press for much more rapid development of renewable energy and energy efficient technologies and cuts in the pollution that causes global warming."[25]

The strange career of *An Inconvenient Truth* has not yet ended.

Acknowledgments

Writing this book has been a long journey filled with un-expected twists and turns. Along the way, I relied upon the support and kindness of colleagues, friends, and family. It is a pleasure to acknowledge some of them here.

I am grateful to librarians and archivists at the University Archives at the University of Illinois (especially Rory Grennan, Harini Rajagopal, and Lindsey Gilroy), the Vanderbilt Television News Archive, the Swarthmore College Peace Collection (especially Wendy Chmielewski and Mary Beth Sigado), the Billy Ireland Cartoon Library and Museum at Ohio State University (especially Susan Liberator, Marilyn Scott, and Lucy Shelton Caswell), the Wisconsin Center for Film and Theater Research, the Margaret Herrick Library at the Academy of Motion Picture Arts and Sciences, the Lyndon B. Johnson Presidential Library, the Special Collections Research Center at the Syracuse University Libraries, the Beinecke Rare Book and Manuscript Library at Yale University, the American Museum of Natural History Research Library, the Gelardin New Media Center at Georgetown University, the Institute Archives and Special Collections at MIT Libraries, the American Heritage Center at the University of Wyoming, and the Oviatt Library at California State University, Northridge. I also wish to thank the staff at the University of Washington Libraries, where I scoured the stacks each summer and completed large chunks of the magazine and other periodical research for this book. Finally, I relied extensively on the Interlibrary Loans service at Trent's Bata Library and owe much to Sharon Bosnell and other staff members for their enormously helpful assistance.

As this book developed, I was fortunate to share my research with university and public audiences. I am grateful to the following colleagues for invitations to present portions of the book-in-progress: Graeme Wynn at the University of British Columbia; James Opp at Carleton University; Marguerite Shaffer at Miami University; William Wylie at the University of Virginia; Mark Madison at the US Fish and Wildlife Service's National Conservation Training Center in West Virginia; and Catriona Sandilands and Katharine Anderson, both at York University. I also thank Alan MacEachern for inviting me to speak at a wonderful environmental history workshop on Prince Edward Island, and Dehlia Hannah for inviting me to speak at an interdisciplinary symposium on visual culture and climate change at the Chemical Heritage Foundation in Philadelphia. I also presented papers at a number of professional conferences, including the meetings of the American Society for Environmental History, the American Studies Association, and the Organization of American Historians. I thank audiences at all these events for their encouragement, stimulation, and thoughtful questions.

This project was supported by generous funding from a Standard Research Grant from the Social Sciences and Humanities Research Council (SSHRC) and an internal SSHRC grant from Trent University. These grants funded archival research and helped defray the steep copyright permission fees for the use of images in this book. In addition, the Trent University Research Fellowship provided crucial time for writing as I worked on revising and completing the book.

Colleagues and students at Trent University have provided tremendous support. In particular, I wish to thank Dimitry Anastakis, Stephen Bocking, Sally Chivers, Caroline Durand, Michael Eamon, Michael Epp, Kate Ervine, Jennine Hurl-Eamon, Carolyn Kay, Liam Mitchell, Marit Munson, Michael Morse, Van Nguyen-Marshall, Kevin Siena, and John Wadland for their friendship and many stimulating conversations. I owe a special thanks to Gavin Fridell for our convivial sabbatical lunches and for lending his critical insight to portions of the manuscript. I also express my gratitude to Trevor Eves and Erin Stewart-Eves for generously helping prepare the frame captures for this book. Finally, I wish to thank Kristoffer Archibald, Jessica Davidson, and Julia Grummitt for their outstanding research assistance.

Friends and colleagues elsewhere have also helped me immensely. I met Matthew Guterl almost twenty years ago in graduate school. Since that time, I have frequently relied on his friendly advice and critical acumen. I thank him for his remarkably perceptive reading of the full manuscript. I also wish to thank other colleagues for their support and conversations over the years, including Thomas Robertson, Adam Rome, Peggy Shaffer, and Michael Smith.

Portions of part 1 of this book first appeared in the essay "Gas Masks, Pogo, and the Ecological Indian: Earth Day and the Visual Politics of American Environmentalism," *American Quarterly* 60 (March 2008): 67–99, © 2008 The American Studies Association. I am grateful to the American Studies Association for permission to reprint sections of the essay. I also wish to thank Curtis Marez (the editor of *American Quarterly* at the time), the managing board of *American Quarterly*, and the anonymous reviewers for their invaluable feedback. Large portions of chapter 1 will soon appear in *Rendering Nature: Animals, Bodies, Places, Politics*, ed. Marguerite S. Shaffer and Phoebe S. K. Young (Philadelphia: University of Pennsylvania Press, 2015). I thank Peggy and Phoebe for inviting me to contribute to that volume, and for their support and encouragement of my work.

I feel very fortunate to have worked once again with the amazing staff at the University of Chicago Press. As editor, Robert Devens provided terrific advice on numerous occasions and played a formative role in the project's development. After Robert took a position with another press, Timothy Mennel enthusiastically stepped in and graced the project with his considerable editorial talents and experience. I am extremely grateful to Tim for his careful reading of the manuscript and for his consistently sage advice. Russell Damian helped me initiate the long process of gathering copyright permissions. Nora Devlin patiently responded to my questions and provided excellent support as I worked on securing all the permissions and illustrations. In the final stages, Renaldo Migaldi's superb copyediting and consummate professionalism helped make this a better book. As readers for the press, Erika Doss and Kathryn Morse provided generous advice, insightful criticism, and thoughtful suggestions that improved this book immeasurably. I want to thank both Erika and Kathy for their exemplary critical engagement.

I am grateful to my family for their love and support, including my parents, Donna and Mike Dunaway; my in-laws, Linda and Peter Capell; and Warren Capell, Sarah Faubel, and Ella Capell.

Above all, I want to thank Dana Capell for everything. She provided boundless support and loving encouragement in every possible way—from reading my prose to sustaining my spirit. I dedicate this book to our children, Max and Zoe, who brighten our lives every day with their curiosity, their zeal, and their radiant sense of wonder. Zoe, who was born on Earth Day (of all days), claims that my book is boring, but she still likes looking at the pictures (especially fig. 3.3). I admire and treasure her irrepressible spirit and her incredible love of learning, books, and language. Max thinks that I should spend more time hiking on trails than researching in libraries, but he nevertheless wrote a clever song about Pogo the Possum (fig. 4.1). His exploratory spirit and incredible passion for knowledge inspire me daily. I will never be able to say how grateful I am to Dana, Max, and Zoe for their love and for their marvelous presence in my life.

Notes

ABBREVIATIONS USED IN NOTES

ACA   Advertising Council Archives, University Archives,
University of Illinois

AMNH-SC American Museum of Natural History Research
Library, Special Collections, New York, NY

BSP   Benjamin Spock Papers, Special Collections Research Center, Syracuse University Libraries

CSPF   *The China Syndrome* production file, Margaret Herrick Library, Academy of Motion Picture Arts and
Sciences, Beverly Hills, CA

MDP   Michael Douglas Papers, 1934–80, Wisconsin Center for Film and Theater Research, Madison, WI

RCP   Rachel Carson Papers, Yale Collection of American
Literature, Beinecke Rare Book and Manuscript
Library, Yale University

SR   SANE Inc. Records (DG 58), Swarthmore College
Peace Collection

VTNA   Vanderbilt Television News Archive, Vanderbilt
University

WKC   Walt Kelly Collection, Billy Ireland Cartoon Library
and Museum, Ohio State University

INTRODUCTION

1.   On historians and visual culture, see also Louis P. Masur,
"'Pictures Have Now Become a Necessity': The Use of Images
in American History Textbooks," *Journal of American History* 84 (March 1998): 1409–24; Michael L. Wilson, "Visual

Culture: A Useful Category of Historical Analysis?" in *The Nineteenth-Century Visual Culture Reader*, ed. Vanessa R. Schwartz and Jeannene M. Przyblyski (New York: Routledge, 2004), 26–33; and James W. Cook, "Seeing the Visual in U.S. History," *Journal of American History* 95 (September 2008): 432–41. For examples of historical scholarship on the visual culture of conservation and environmentalism, see Gregg Mitman, *Reel Nature: America's Romance with Wildlife on Film* (Cambridge, MA: Harvard University Press, 1999); and Finis Dunaway, *Natural Visions: The Power of Images in American Environmental Reform* (Chicago: University of Chicago Press, 2005). Scholars in environmental communication, the environmental humanities, and other interdisciplinary fields have also explored environmentalism in visual and popular culture. For some examples, see Kevin Michael DeLuca, *Image Politics: The New Rhetoric of Environmental Activism* (New York: Guilford Press, 1999); David Ingram, *Green Screen: Environmentalism and Hollywood Cinema* (Exeter, UK: University of Exeter Press, 2000); Noël Sturgeon, *Environmentalism in Popular Culture: Gender, Race, Sexuality, and the Politics of the Natural* (Tucson: University of Arizona Press, 2009); and Adrian J. Ivakhiv, *Ecologies of the Moving Image: Cinema, Affect, Nature* (Waterloo, Canada: Wilfrid Laurier University Press, 2013).

2. My argument here builds on the work of scholars who have revised Jurgen Habermas's theory of the public sphere by examining the emotionality of public culture and explaining how the public is constituted through acts of common spectatorship. See Michael Warner, *Publics and Counterpublics* (New York: Zone Books, 2002); and Robert Hariman and John Louis Lucaites, *No Caption Needed: Iconic Photographs, Public Culture, and Liberal Democracy* (Chicago: University of Chicago Press, 2007). For an overview of theoretical approaches to public culture, see Mary Kupiec Cayton, "What Is Public Culture? Agency and Contested Meaning in American Culture—An Introduction," in *Public Culture: Diversity, Democracy, and Community in the United States*, ed. Marguerite S. Shaffer (Philadelphia: University of Pennsylvania Press, 2008), 1–25. On visual culture, affect, and the public sphere, see Erika Doss, *Memorial Mania: Public Feeling in America* (Chicago: University of Chicago Press, 2010); and Erika Doss, "Makes Me Laugh, Makes Me Cry: Feelings and American Art," *American Art* 25 (Fall 2011): 2–8. On the links between emotions and rational thought, see Martha C. Nussbaum, *Upheavals of Thought: The Intelligence of Emotions* (New York: Cambridge University Press, 2001); and George E. Marcus, *The Sentimental Citizen: Emotion in Democratic Politics* (University Park: Pennsylvania State University Press, 2002).

3. Often discussed by philosophers and political theorists, environmental citizenship has rarely found its way into historical accounts of US environmentalism. Representative works by political theorists and other scholars include Andrew Dobson, *Citizenship and the Environment* (New York:

Oxford University Press, 2003); and *Environmental Citizenship*, ed. Andrew Dobson and Derek Bell (Cambridge, MA: MIT Press, 2006). In addition to these works, the historian Michelle Murphy has defined biocitizenship as "a useful term naming efforts that take life—from human bodies to ecosystems—as points of entry into making demands of the state, and thereby articulating the terms of citizenship via health and living-being." See Michelle Murphy, "Chemical Regimes of Living," *Environmental History* 13 (October 2008): 699. Murphy's article builds on the work of the anthropologist Adriana Petryna, *Life Exposed: Biological Citizens after Chernobyl* (Princeton, NJ: Princeton University Press, 2002).

4. For critiques of the discourse of individual responsibility, see also Ted Steinberg, "Can Capitalism Save the Planet? On the Origins of Green Liberalism," *Radical History Review* 107 (Spring 2010): 7–24; Michael F. Maniates, "Individualization: Plant a Tree, Buy a Bike, Save the World?," *Global Environmental Politics* 1 (August 2001): 31–52; and Timothy W. Luke, "Green Consumerism: Ecology and the Ruse of Recycling," in *In the Nature of Things: Language, Politics, and the Environment*, ed. Jane Bennett and William Chaloupka (Minneapolis: University of Minnesota Press, 1993), 154–72.

5. Scholars of civil rights, feminism, antiwar, and other social movements have also examined the limits of media representation and explained how popular media sources marginalized or trivialized radical political ideas. For examples, see Martin A. Berger, *Seeing through Race: A Reinterpretation of Civil Rights Photography* (Berkeley: University of California Press, 2011); Susan J. Douglas, *Where the Girls Are: Growing Up Female with the Mass Media* (New York: Times Books, 1994); and Todd Gitlin, *The Whole World Is Watching: Mass Media in the Making and Unmaking of the New Left* (Berkeley: University of California Press, 1980).

6. Rob Nixon, *Slow Violence and the Environmentalism of the Poor* (Cambridge, MA: Harvard University Press, 2011), 2.

7. For critiques of environmentalism, see, among others, Robert Gottlieb, *Forcing the Spring: The Transformation of the American Environmental Movement* (Washington: Island Press, 1993).

8. Slavoj Žižek, *Violence* (New York: Picador, 2008), 2.

9. I borrow the term "environmental moment" from *The Environmental Moment, 1968–1972*, ed. David Stradling (Seattle: University of Washington Press, 2012).

CHAPTER ONE

1. SANE, advertisement, *New York Times*, April 16, 1962.

2. Milton S. Katz, *Ban the Bomb: A History of SANE, the Committee for a Sane Nuclear Policy* (New York: Praeger, 1987), 75; and Benjamin Spock, *Baby and Child Care*, rev. ed. (New York: Pocket Books, 1957), 3–4.

3. For an important critique of the wilderness aesthetic, see William Cronon, "The Trouble with Wilderness; or, Getting Back to the Wrong Nature," in *Uncommon Ground: Toward Reinventing Nature*, ed. William Cronon (New York: W. W. Norton, 1995), 69–90. On Sierra Club photography during the 1960s, see Finis Dunaway, *Natural Visions: The Power of Images in American Environmental Reform* (Chicago: University of Chicago Press, 2005), chaps. 5–7.

4. On the atomic sublime and mushroom cloud imagery as political pacification, see Peter B. Hales, "The Atomic Sublime," *American Studies* 32 (Spring 1991): 5–31; Scott Kirsch, "Watching the Bombs Go Off: Photography, Nuclear Landscapes, and Spectator Democracy," *Antipode* 29 (July 1997): 227–55; Peter Bacon Hales, "Imagining the Atomic Age: *Life* and the Atom," in *Looking at Life Magazine*, ed. Erika Doss (Washington: Smithsonian Institution Press, 2001), 103–19; and Robert Hariman and John Louis Lucaites, "The Iconic Image of the Mushroom Cloud and the Cold War Nuclear Optic," in *Picturing Atrocity: Photography in Crisis*, ed. Geoffrey Batchen, Mick Gidley, Nancy K. Miller, and Jay Prosser (London: Reaktion Books, 2012), 135–45.

5. This military film clip is included in *The Atomic Cafe: Collector's Edition*, DVD, directed by Jayne Loader, Kevin Rafferty, and Pierce Rafferty (1982; New York: New Video Group, 2008). On the ecological body, see Maril Hazlett, "Voices from the *Spring*: *Silent Spring* and the Ecological Turn in American Health," in *Seeing Nature through Gender*, ed. Virginia J. Scharff (Lawrence: University Press of Kansas, 2003), 103–28; and Linda Nash, *Inescapable Ecologies: A History of the Environment, Disease, and Knowledge* (Berkeley: University of California Press, 2006).

6. For background on *The Big Picture*, see Stanley Field, "The Big Picture," *Journal of Broadcasting and Electronic Media* 6 (Spring 1962): 125–27; J. Fred MacDonald, "The Cold War as Entertainment in 'Fifties Television," *Journal of Popular Film and Television* 7 (Spring 1978): 3–31, esp. 14; and Nancy E. Bernhard, *U.S. Television News and Cold War Propaganda, 1947–1960* (New York: Cambridge University Press, 1999), 142–43. The episode that I discuss can be found at http://video.google.com /videoplay?docid=-2149878949626814717#.

7. SANE, advertisement, *New York Times*, November 15, 1957. Arno G. Huth, "Response to the First Statement issued by the National Committee for a Sane Nuclear Policy, November 15 to December 31, 1957: A Preliminary Analysis," January 1958, in subseries B-1, box 4, folder "SANE, National Office—SANE Files of Norman Cousins, 1957–1958, Huth, Arno," SANE Inc. Records (DG 58), Swarthmore College Peace Collection (hereafter SR), quotations from 13 and 14.

8. SANE, advertisement, *New York Times*, April 11, 1958. On different conceptions of environmental apocalypse, see also Frederick Buell, *From Apoca-*

lypse to Way of Life: Environmental Crisis in the American Century (New York: Routledge, 2004).

9. Wilma Holden to Norman Cousins, April 24, 1958, box 2, folder "Correspondence of Norman Cousins (as editor of *The Saturday Review*), 'H,' 1958," series B-1, SR. On images, emotions, and democratic citizenship, see Robert Hariman and John Louis Lucaites, *No Caption Needed: Iconic Photographs, Public Culture, and Liberal Democracy* (Chicago: University of Chicago Press, 2007).

10. Lauren Berlant, *The Queen of America Goes to Washington City: Essays on Sex and Citizenship* (Durham: Duke University Press, 1997), 5.

11. SANE, advertisement, *New York Times*, April 16, 1962.

12. Thomas Frank, *The Conquest of Cool: Business Culture, Counterculture, and the Rise of Hip Consumerism* (Chicago: University of Chicago Press, 1997), chap. 3.

13. First quotation from an unnamed letter writer quoted in Nell Lee Litvak, public information director of SANE, form letter to SANE members, April 21, 1962, copy in box 54, folder "Politics, SANE, Memoranda, Gen-1963," Benjamin Spock Papers, Special Collections Research Center, Syracuse University Libraries (hereafter BSP); second quotation from Mary M. Grooms to Homer Jack, April 21, 1962, copy in box 10, folder "Correspondence, April 17, 1962–April 24, 1962," BSP; third quotation from Mrs. John T. McClure to National Committee for a Sane Nuclear Policy, April 28, 1962, copy in box 10, folder "Correspondence, April 25, 1962–April 20, 1962," BSP.

14. Jeanne S. Bagby to Benjamin Spock, April 20, 1962, in box 10, folder "Correspondence, April 17, 1962–April 24, 1962," BSP. On the politics of gender and antinuclear protest during this period, see Amy Swerdlow, *Women Strike for Peace: Traditional Motherhood and Radical Politics in the 1960s* (Chicago: University of Chicago Press, 1993); and Lawrence S. Wittner, "Gender Roles and Nuclear Disarmament Activism, 1954–1965," *Gender and History* 12 (April 2000): 197–222.

15. "SANE—and Others," *Time*, April 27, 1962, 22–23 (quotation on 22).

16. Dentists for SANE, advertisement, *New York Times*, April 7, 1963. On Barry Commoner and the Committee for Nuclear Information, see Michael Egan, *Barry Commoner and the Science of Survival: The Remaking of American Environmentalism* (Cambridge, MA: MIT Press, 2007), chap. 2; and Kelly Moore, *Disrupting Science: Social Movements, American Scientists, and the Politics of the Military, 1945–1975* (Princeton, NJ: Princeton University Press, 2008), chap. 4.

17. Dentists for SANE, advertisement, *New York Times*, April 7, 1963.

18. On popular depictions of suburban communities, see, for example, Wendy Kozol, *Life's America: Family and Nation in Postwar Photojournalism* (Philadelphia: Temple University Press, 1994).

19. For background on this commercial and the 1964 campaign, see Kathleen Hall Jamieson, *Packaging the Presidency: A History and Criticism of Presidential Campaign Advertising*, 2nd ed. (New York: Oxford University Press, 1992), 198–202; and Edwin Diamond and Stephen Bates, *The Spot: The Rise of Political Advertising on Television* (Cambridge, MA: MIT Press, 1984), 127–34. The commercial can be viewed at, among other sites, Living Room Candidate, a website maintained by the Museum of the Moving Image: http://www.livingroomcandidate.org/commercials/1964.

20. This commercial can be viewed at http://www.livingroomcandidate.org /commercials/1964. For background on the commercial, see Jamieson, 201.

21. Adlai Stevenson, as quoted in Egan, *Barry Commoner and the Science of Survival*, 52; Lyndon B. Johnson, as quoted in Egan, 75.

22. On these issues, see also Paul Boyer, "From Activism to Apathy: The American People and Nuclear Weapons, 1963–1980," *Journal of American History* 70 (March 1984): 821–44; and Kirsch, esp. 246–48, on how the treaty helped obscure ongoing health and environmental problems related to the arms race.

23. *CBS Reports*, "The Silent Spring of Rachel Carson," April 3, 1963, DVD available from CBS News Archives, New York. All quotations and visual descriptions of this episode are based on my viewing of this DVD. Carson read excerpts from *Silent Spring* during the episode, including the quotation in this paragraph from Carson, *Silent Spring* (Boston: Houghton Mifflin, 1962), 7–8.

24. F. Barrows Colton, "Your New World of Tomorrow," *National Geographic*, October 1945, 385–410 (photograph on 410); "Fogging," *Life*, July 19, 1948, 49–51 (photograph on 51).

25. "Pesticides: The Price for Progress," *Time*, September 28, 1962, 45–48 (quotations on 45 and 48). On the gendered response to *Silent Spring*, see Michael B. Smith, "'Silence, Miss Carson!': Science, Gender, and the Reception of *Silent Spring*," *Feminist Studies* 27 (Fall 2001): 733–72; Hazlett, "Voices from the *Spring*"; Hazlett, "'Woman vs. Man vs. Bugs': Gender and Popular Ecology in Early Reactions to *Silent Spring*," *Environmental History* 9 (October 2004): 701–29; and Linda Lear, *Rachel Carson: Witness for Nature* (New York: Henry Holt, 1997). On gender and environmental politics during this period, see also Adam Rome, "'Give Earth a Chance': The Environmental Movement and the Sixties," *Journal of American History* 90 (September 2003): 525–54, esp. 534–41; and Nancy C. Unger, *Beyond Nature's Housekeepers: American Women in Environmental History* (New York: Oxford University Press, 2012), chap. 6.

26. On *CBS Reports* and other news documentaries during this period, see Michael Curtin, *Redeeming the Wasteland: Television Documentary and Cold War Politics* (New Brunswick, NJ: Rutgers University Press, 1995). Warner Twyford, "3 Sponsors Won't Sing of Spring," *Virginia Pilot*, April 3, 1963, copy in box 75, folder 1335, Rachel Carson Papers, Yale Collection of

American Literature, Beinecke Rare Book and Manuscript Library, Yale University (hereafter RCP). On media framings of *Silent Spring*, including an analysis of *CBS Reports*, see also Gary Kroll, "The 'Silent Springs' of Rachel Carson: Mass Media and the Origins of Modern Environmentalism," *Public Understanding of Science* 10 (2001): 403–20.

27. Patricia Costa, "CBS Looks at Foggy Issue: Problem of Pesticide Control," *Rochester Democrat and Chronicle*, April 4, 1963, copy in box 75, folder 1336, RCP. See also Frank T. Adams, Jr., "Spring Fever," *Virginian Pilot*, April 7, 1963, copy in box 75, folder 1337, RCP.

28. For more on the "mad scientist" image, see Christopher Frayling, *Mad, Bad and Dangerous? The Scientist and the Cinema* (London: Reaktion Books, 2005); and David J. Skal, *Screams of Reason: Mad Science and Modern Culture* (New York: W. W. Norton & Company, 1998). The quotation is from Frayling, 36.

29. Adams, "Spring Fever"; "TV Round-Up on 'Silent Spring,'" *Fitchburg Sentinel*, April 13, 1963, copy in box 75, folder 1336, RCP.

30. Gordon Young, "Pollution, Threat to Man's Only Home," *National Geographic*, December 1970, 758; the photograph by James P. Blair appeared on 759. Don Moser, "A Lament for Some Companions of My Youth," *Life*, January 22, 1971, 52; the photograph by George Silk also appeared on 52.

31. Lynne W. Heutchy, letter to editor, *Life*, February 12, 1971, 14A.

32. Environmental Defense Fund, advertisement, *New York Times*, March 29, 1970; Environmental Defense Fund, advertisement, *Audubon*, November 1970, 109.

33. Wilfrid Sheed, "Evolution on a Bad Trip," *Life*, July 11, 1969, 7; the photograph by Rowland Scherman also appeared on p. 7. James A. Oliver, "Can Man Survive?," *Parents' Magazine and Better Family Living*, August 1969, 40.

34. These comments are based on my viewing of the following news broadcasts: CBS News, December 27, 1968; ABC News, May 19, 1969; CBS News, October 27, 1969; ABC News, November 12, 1969; CBS News, November 12, 1969; NBC News, September 2, 1970; CBS News, June 14, 1972; and NBC News, June 14, 1972, all available from Vanderbilt Television News Archive, Vanderbilt University (hereafter VTNA). The Sevareid quotation is from CBS News, November 12, 1969. On the decision to ban DDT, see Thomas R. Dunlap, *DDT: Scientists, Citizens, and Public Policy* (Princeton, NJ: Princeton University Press, 1981), 231–45.

CHAPTER TWO

1. ABC News, February 5, 1969, available from VTNA.

2. For background on the spill, see Robert Easton, *Black Tide: The Santa Barbara Oil Spill and Its Consequences* (New York: Delacorte Press, 1972).

3. "California: The Great Blob," *Newsweek*, February 17, 1969, 31–32 (quotations on 31). For more on the Cuyahoga fire, including a perceptive

discussion of its subsequent emergence as an icon of environmental crisis, see David Stradling and Richard Stradling, "Perceptions of the Burning River: Deindustrialization and Cleveland's Cuyahoga River," *Environmental History* 13 (July 2008): 515–35.

4. Archibald MacLeish, "A Reflection: Riders on Earth Together, Brothers in Eternal Cold," *New York Times*, December 25, 1968. For more on the significance of *Earthrise* as well as the 1972 whole earth photograph, see Denis Cosgrove, "Contested Global Visions: *One-World, Whole-Earth,* and the Apollo Space Photographs," *Annals of the Association of American Geographers* 84 (June 1994): 270–94; Neil Maher, "Shooting the Moon," *Environmental History* 9 (July 2004): 526–31; and Robert Poole, *Earthrise: How Man First Saw the Earth* (New Haven: Yale University Press, 2008). Walter A. Kleinschrod, letter, *New York Times*, January 5, 1969; *The Last Whole Earth Catalog,* August 1972, 1. *Earthrise* also appeared on the cover of *Whole Earth Catalog,* Spring 1969. On the significance of the *Whole Earth Catalog* to environmentalism, see Andrew G. Kirk, *Counterculture Green: The Whole Earth Catalog and American Environmentalism* (Lawrence: University Press of Kansas, 2007).

5. Gaylord Nelson to Frank Stanton, April 7, 1971, box 2, folder 15, Gaylord Nelson Collection mss1020, Wisconsin Historical Society, Madison, WI, available at http://www.nelsonearthday.net/collection/beginning-cbsnews-letter.htm. See also Adam Rome, *The Genius of Earth Day: How a 1970 Teach-In Unexpectedly Made the First Green Generation* (New York: Hill and Wang, 2013), 57–58 and 67–68; and Bill Christofferson, *The Man from Clear Lake: Earth Day Founder Senator Gaylord Nelson* (Madison: University of Wisconsin Press, 2004), 302.

6. "Earth Day and Space Day," *New York Times*, April 19, 1970; Margaret Mead, "Earth People," address given at Bryant Park, New York City, April 22, 1970, reprinted in *Earth Day—The Beginning: A Guide for Survival,* ed. Environmental Action (New York: Bantam Books, 1970), 223.

7. My comments here have been influenced by Maher, "Shooting the Moon."

8. The caption quoted is from Ross Macdonald and Robert Easton, "Santa Barbarans Cite an 11th Commandment: 'Thou Shalt Not Abuse the Earth,'" *New York Times Magazine*, October 12, 1969, 32. The second quotation is from Macdonald and Easton, 32.

9. David Snell, "Iridescent Gift of Death," *Life*, June 13, 1969, 22–27 (quotations on 23).

10. Snell, "Iridescent Gift of Death," 26.

11. The remark attributed to Hartley appeared in, among other sources, Warren Weaver Jr., "The Oil Threat to the Beaches," *New York Times*, February 9, 1969.

12. The first quotations are from Union Oil Company, advertisement, *New York Times*, February 20, 1969. Versions of Hartley's letter also appeared in Fred L. Hartley, letter to the editor, *Wall Street Journal*, February 14, 1969;

and Fred L. Hartley, letter to the editor, *Time*, February 21, 1969, 5. The final quotation is from "Head of Union Oil Denies Remarks on Loss of Birds in Slick," *New York Times*, February 19, 1969.

13. Harvey Molotch, "Santa Barbara: Oil in the Velvet Playground," *Ramparts*, November 1969, 43–51 (quotation on 50). As Kathryn Morse explains, in the period just before the Santa Barbara spill, media reports of the *Torrey Canyon* and other oil disasters also began to link these spills to a broader sense of ecological crisis. See Kathryn Morse, "There Will Be Birds: Images of Oil Disasters in the Nineteenth and Twentieth Centuries," *Journal of American History* 99 (June 2012): 124–34, esp. 129.

14. Terrence Kester, letter to the editor, *Life*, July 4, 1969, 16A.

15. Gladwin Hill, "One Year Later, Impact of Great Oil Slick Is Still Felt," *New York Times*, January 25, 1970; Edgar V. Roberts, letter to editor, *New York Times*, February 23, 1969.

16. Paul Sabin, "Crisis and Continuity in U.S. Oil Politics, 1965–1980," *Journal of American History* 99 (June 2012): 177–86 (quotations on 181).

17. *National Geographic*, December 1970, cover. The special issue was titled "Our Ecological Crisis." The photograph by Bruce Dale appeared on the cover and on 755. Roland Barthes, *Camera Lucida: Reflections on Photography*, trans. Richard Howard (New York: Hill and Wang, 1981), 97. Alexander Wetmore quoted in Gordon Young, "Pollution, Threat to Man's Only Home," *National Geographic*, December 1970, 754.

18. The first quotation is from the title of the article by Young, "Pollution, Threat to Man's Only Home," which begins on 738; the second quotation is also from 738.

CHAPTER THREE

1. Ralph Graves, "Editor's Note: Why John Pekkanen Gave Up Eating Liver," *Life*, January 30, 1970, 3.

2. Ray Osrin's "The Thinker" first appeared in the Cleveland *Plain Dealer* on January 11, 1970. It was reproduced as a poster for Earth Day and reprinted in several publications, including Luther J. Carter, "Earth Day: A Fresh Way of Perceiving the Environment," *Science*, May 1, 1970, 558. The gas mask poster of Christ on the cross is reprinted in Gary Yanker, *Prop Art: Over 1000 Contemporary Political Posters* (New York: Darien House, 1972), 203. The illustration of Alfred E. Neuman with a gas mask appeared in *Mad*, October 1971, 25. Many examples of the portrayal of gas masks in television news coverage of Earth Day can be seen on broadcasts from April 22, 1970, and April 23, 1970, VTNA.

3. T. George Harris, "Editorial: Dirty Pictures," *Psychology Today*, March 1970, 26 (first quotation); Scot Morris, "Dirty Pictures: The Winners," *Psychology Today*, November 1970, 94–102 (quotations on 94 and 102), artwork on 95.

4. Morris Neiburger quoted in "Professor Says World Will Smother in Smog," *New York Times*, August 9, 1965. See also Scott Hamilton Dewey, *Don't Breathe the Air: Air Pollution and U.S. Environmental Politics, 1945–1970* (College Station: Texas A&M Press, 2000), 105.

5. On middle-class white women and air pollution activism, see Adam Rome, *The Genius of Earth Day: How a 1970 Teach-In Unexpectedly Made the First Green Generation* (New York: Hill and Wang, 2013), 30–31 and 36–37; and Nancy C. Unger, *Beyond Nature's Housekeepers: American Women in Environmental History* (New York: Oxford University Press, 2012), 144.

6. Modris Eksteins, *Rites of Spring: The Great War and the Birth of the Modern Age* (New York: Anchor Books/Doubleday, 1990), 163. Hugh R. Slotten, "Humane Chemistry or Scientific Barbarism? American Responses to World War I Poison Gas, 1915–1930," *Journal of American History* 77 (September 1990): 476–98; and Edmund Russell, *War and Nature: Fighting Humans and Insects with Chemicals from World War I to Silent Spring* (New York: Cambridge University Press, 2001), especially 90–91.

7. On the Smog-a-Tears, see Dewey, *Don't Breathe the Air*, 93–94; and Unger, *Beyond Nature's Housekeepers*, 144. On air pollution activism in Pasadena and other southern California suburbs, see Christopher C. Sellers, *Crabgrass Crucible: Suburban Nature and the Rise of Environmentalism in Twentieth-Century America* (Chapel Hill: University of North Carolina Press, 2012), chap. 7.

8. Tom Lehrer, "Pollution," (1965), lyrics reprinted in *The Sierra Club Survival Songbook*, ed. Jim Morse and Nancy Mathews (San Francisco: Sierra Club, 1971), 18–21. John Gardner, quoted in Dewey, *Don't Breathe the Air*, 249. Barry Commoner, *Science and Survival* (New York: Viking Press, 1967), cover.

9. "Fighting to Save the Earth from Man," *Time*, February 2, 1970, 63. For another example of this type of poster, see Yanker, *Prop Art*, 204. The last quotation is from Maria Warner, *Alone of All Her Sex: The Myth and the Cult of the Virgin Mary* (New York: Knopf, 1976), 198.

10. See Wendy Kozol, "Madonnas of the Fields: Photography, Gender, and 1930s Farm Relief," *Genders* 2 (1988): 1–23; Laura Briggs, "Mother, Child, Race, Nation: The Visual Iconography of Rescue and the Politics of Transnational and Transracial Adoption," *Gender and History* 15 (August 2003): 179–200; and Robert Hariman and John Louis Lucaites, *No Caption Needed: Iconic Photographs, Public Culture, and Liberal Democracy* (Chicago: University of Chicago Press, 2007), 53–67.

11. Michelle Murphy, "Uncertain Exposures and the Privilege of Imperception: Activist Scientists and Race at the U.S. Environmental Protection Agency," *Osiris* 19 (2004): 266–82. On perception and imperception in the history of environmental injustice, see also Linda Nash, "The Fruits of Ill-Health: Pesticides and Workers' Bodies in Post–World War II California," *Osiris* 19 (2004): 203–19; and Gregg Mitman, *Breathing Space: How Allergies*

Shape Our Lives and Landscapes (New Haven: Yale University Press, 2007), chap. 4. The Commoner quotations are from Barry Commoner, *The Closing Circle: Nature, Man, and Technology* (New York: Knopf, 1971), 208. Although beyond the scope of this chapter, it is worth nothing that gas masks would be used by environmental justice activists in the 1980s and 1990s in campaigns that emphasized the disproportionate levels of exposure to air pollutants among racial minorities. See Julie Sze, *Noxious New York: The Racial Politics of Urban Health and Environmental Justice* (Cambridge, MA: MIT Press, 2007), 99–100; and Mitman, *Breathing Space*, 164.

12. On Earth Day organizers' efforts to link environmental issues with social justice, see Rome, *The Genius of Earth Day*, 85–97. Denis Hayes is quoted on 86. The welfare rights activist Jeannette Washington is quoted in "Action Notes," *Environmental Action: April 22*, March 12, 1970, 3. See also "Lead Poisons Ghetto Children," *Environmental Action: April 22*, April 2, 1970, 3.

13. See Robert Gottlieb, *Forcing the Spring: The Transformation of the American Environmental Movement* (Washington: Island Press, 1993), 244–50.

14. Jack Newfield, "Lead Poisoning: Silent Epidemic in the Slums," *Village Voice*, September 18, 1969, reprinted in *The Environmental Moment, 1968–1972*, ed. David Stradling (Seattle: University of Washington Press, 2012), 43–48 (quotation on 45).

15. NBC News, April 23, 1970, VTNA; Jack Rosenthal, "Some Troubled by Environment Drive," *New York Times*, April 22, 1970. Moreover, as the historian Christopher C. Sellers has shown, when some African American protesters wore gas masks on Earth Day, they were "effectively white-washed." In one newspaper photograph, for example, cropping effaced their presence and eliminated the "civil rights message of [a] protest sign." See Sellers, 281.

16. For coverage of Black Survival in St. Louis newspapers, see "Many Events Scheduled Here for Environmental Observation," *St. Louis Post-Dispatch*, April 19, 1970, and "'Black Survival' Group Pushes Pollution Fight," *St. Louis Post-Dispatch*, April 26, 1970. On Ivory Perry's involvement with this struggle, see George Lipsitz, *A Life in the Struggle: Ivory Perry and the Culture of Opposition*, rev. ed. (Philadelphia: Temple University Press, 1995), chap. 7. For the Earth Day skits, see Freddie Mae Brown and the St. Louis Metropolitan Black Survival Committee, "Black Survival: A Collage of Skits," in *Earth Day—The Beginning: A Guide for Survival*, ed. Environmental Action (New York: Bantam Books, 1970), 95–105.

17. Laura Pulido, *Environmentalism and Economic Justice: Two Chicano Struggles in the Southwest* (Tucson: University of Arizona Press, 1996), xv. On the importance of positionality to both mainstream and subaltern environmentalisms, see Pulido, 25–30. Ulrich Beck, *Risk Society: Toward a New Modernity*, trans. Mark Ritter (London: Sage Publications, 1992), 74.

18. Nathan Hare, "Black Ecology," *Black Scholar* 1 (April 1970): 5; Jack New-field, "Let Them Eat Lead," *New York Times*, June 16, 1971. For more on Jack Newfield's efforts to publicize this issue, see Christian Warren, *Brush with Death: A Social History of Lead Poisoning* (Baltimore: Johns Hopkins University Press, 2000), chap. 10.

19. On the UFWOC, see Pulido, *Environmentalism and Economic Justice*, chap. 3; Robert Gordon, "Poisons in the Fields: The United Farm Workers, Pesticides, and Environmental Politics," *Pacific Historical Review* 68 (February 1999): 51–77; Nash, "The Fruits of Ill-Health"; and Linda Nash, *Inescapable Ecologies: A History of Environment, Disease, and Knowledge* (Berkeley: University of California Press, 2006), 161–67. See also "Mothers Alarmed at DDT Danger," *El Malcriado*, August 1–15, 1969, 3, 11, which includes a photograph of nursing mothers protesting the use of DDT.

20. See Gottlieb, *Forcing the Spring*, chap. 7.

21. Sam Love, "Ecology and Social Justice: Is There a Conflict?" *Environmental Action*, August 5, 1972, 5 (both quotations).

22. Andrew Hurley, *Environmental Inequalities: Class, Race, and Industrial Pollution in Gary, Indiana, 1945–1980* (Chapel Hill: University of North Carolina Press, 1995), 162; George Lipsitz, "Toxic Racism," *American Quarterly* 47 (September 1995): 420.

23. Many photographs of men wearing gas masks appeared during this period. See, for example, "Fighting to Save the Earth from Man," 62.

24. Timothy Dumas, "Unmasked," *Smithsonian*, July/August 2010, 8–10; all quotations on 10.

CHAPTER FOUR

1. Roy Alexander to Walt Kelly, February 5, 1970, in box 14, folder 14, Walt Kelly Collection, Billy Ireland Cartoon Library & Museum, Ohio State University (hereafter WKC). Delivered on January 19, 1970, Nelson's speech is printed in *Congressional Record*, 91st Cong., 2d sess., 81–85 (Pogo quotation on 82).

2. Jack L. Alford to Walt Kelly, February 6, 1970, in box 14, folder 14, WKC.

3. This poster is reprinted in Selby Kelly and Steve A. Thompson, *Pogo Files for Pogophiles: A Retrospective on 50 Years of Walt Kelly's Classic Comic Strip* (Richfield, MN: Spring Hollow Books, 1992), n.p.

4. On "governing at a distance," see Nikolas Rose and Peter Miller, "Political Power beyond the State: Problematics of Government," *British Journal of Sociology* 43 (June 1992): 173–205; and Peter Miller and Nikolas Rose, *Governing the Present: Administering Economic, Social and Personal Life* (Malden, MA: Polity Press, 2008), 33–34 and passim.

5. Lucy Shelton Caswell, introduction to *Walt Kelly: A Retrospective Exhibition to Celebrate the Seventy-fifth Anniversary of His Birth* (Columbus: Ohio State

University Libraries, 1988), 10. See also Steve Thompson, "Highlights of Pogo," in *Walt Kelly: A Retrospective*, 45. On the relationship between word and image in the comics, see also Robert C. Harvey, *The Art of the Funnies: An Aesthetic History* (Jackson: University Press of Mississippi, 1994), 9–10.

6. Sam Love quoted in Peter Harnik, "Debunking Madison Avenue," *Environmental Action*, September 4, 1971, 10–11; quotation on 11.

7. For background on Walt Kelly's career, see Stephen E. Kercher, *Revel with a Cause: Liberal Satire in Postwar America* (Chicago: University of Chicago Press, 2006), 57–74 (quotation on 64); and Harvey, *The Art of the Funnies*, 185–201.

8. For Kelly's explanation of the origin of the quote, see Walt Kelly, "Zeroing In on Those Polluters: We Have Met the Enemy and He Is Us," in *The Best of Pogo*, ed. Selby Kelly and Bill Crouch (New York: Simon and Schuster, 1982), 224.

9. Robert M. Smith, "Museum Uses Psychedelic Lights and Electronic Music to Show That Life Can Be Ugly," *New York Times*, May 19, 1969. On *Can Man Survive?*, see also Alison Griffiths, *Shivers Down Your Spine: Cinema, Museums, and the Immersive View* (New York: Columbia University Press, 2008), 252–64.

10. Wilfrid Sheed, "Evolution on a Bad Trip," *Life*, July 11, 1969, 7.

11. I found the following photograph (taken in 1969) of this population graph with the Pogo quotation: neg. no. 61814–11A:160, in American Museum of Natural History Research Library, Special Collections, New York (hereafter AMNH-SC).

12. The "Population Explosion" poster was reprinted in *Environmental Action*, March 12, 1970, 2; Paul R. Ehrlich, *The Population Bomb* (New York: Ballantine Books, 1968), 66–67. On Ehrlich and the broader concern about overpopulation in postwar America, see Thomas Robertson, *The Malthusian Moment: Global Population Growth and the Birth of American Environmentalism* (New Brunswick, NJ: Rutgers University Press, 2012).

13. "Exhibit Asks Grim Query," *Newsday*, May 16, 1969; and Jane Allison, "We Have Met the Enemy and They Are Us: Pogo," *Indianapolis News*, May 28, 1969, copies in Vertical File, *Can Man Survive?*, AMNH-SC.

14. Patricia Welbourn, "Can Man Survive?" *Victoria Daily Times Weekend Magazine*, January 3, 1970, 5–7 (quotation on 5), copy in Vertical File, *Can Man Survive?*, AMNH-SC.

15. Kenneth Auchincloss, "The Ravaged Environment," *Newsweek*, January 20, 1970, 32.

16. The first photograph and caption appeared in ibid. The collage was featured in the same issue of *Newsweek* as part of a color gallery, which was not paginated. The caption appeared after the gallery on 37.

17. John Lear, "The Enemy Is Us," *Saturday Review*, March 7, 1970, 58. For uses of Pogo's quote at Earth Day events see Joseph Lelyveld, "Mood Is

Joyful as City Gives Its Support," *New York Times*, April 23, 1970; and Karl E. Meyer, "Joy in New York—Autoless Streets," *Washington Post*, April 23, 1970. The book review mentioned is Christopher Lehmann-Haupt, "Troubles aboard the Spaceship Earth," *New York Times*, April 22, 1970.

18. Wallace Stegner, "'We Have Met the Enemy, and He Is Us,'" *Life*, July 10, 1970, 10–11.

19. The photograph and caption appeared in Gordon Young, "Pollution, Threat to Man's Only Home," *National Geographic*, December 1970, 750. Pogo was quoted in the same article on 748.

20. On environmental justice struggles in New York City during the 1960s and beyond, see Matthew Gandy, *Concrete and Clay: Reworking Nature in New York City* (Cambridge, MA: MIT Press, 2002); Gregg Mitman, *Breathing Space: How Allergies Shape Our Lives and Landscapes* (New Haven: Yale University Press, 2007), chap. 4; and Julie Sze, *Noxious New York: The Racial Politics of Urban Health and Environmental Justice* (Cambridge, MA: MIT Press, 2007).

21. On neoliberal models of citizenship, see, for example, Nikolas Rose, "Governing 'Advanced' Liberal Democracies," in *Foucault and Political Reason: Liberalism, Neoliberalism, and Rationalities of Government*, ed. Andrew Barry, Thomas Osborne, and Nikolas Rose (Chicago: University of Chicago Press, 1996), 37–64; and Laurie Ouellette, "'Take Responsibility for Yourself': *Judge Judy* and the Neoliberal Citizen," in *Reality TV: Remaking Television Culture*, ed. Susan Murray and Laurie Ouellette (New York: New York University Press, 2004), 231–50.

22. Both strips have been reprinted many times. The first, for example, can be seen in *The Best of Pogo*, 163, while the second is in Bruce Kennedy, "Have We Really Met the Enemy Yet?" *Environmental Action*, October 2, 1971, 13–14; strip on 13.

23. Walt Kelly, "The Many Faceted Public Eye," in his *Pogo: We Have Met the Enemy and He Is Us* (New York: Simon and Schuster, 1972), 10.

24. *Earth Day—The Beginning: A Guide for Survival*, ed. Environmental Action (New York: Bantam Books, 1970), 231.

25. Gene Marine, "Scorecard on the Environment," *Ramparts*, December 1973, 18–20; quotations on 20.

26. Kennedy, "Have We Really Met the Enemy Yet?," 13. According to *Environmental Action*, the article originally appeared in the July 1971 issue of *Bonanza*, a newsletter published by a Sierra Club chapter in Sacramento, California.

27. Kennedy, 13–14.

28. Robert C. Harvey, *Children of the Yellow Kid: The Evolution of the American Comic Strip* (Seattle: Frye Art Museum in association with the University of Washington Press, 1998), 135.

CHAPTER FIVE

1. Robert Schnakenberg, "Keep America Beautiful, Inc.: Crying Indian Campaign," in *Encyclopedia of Major Marketing Campaigns*, ed. Thomas Riggs (Detroit: Thomson Gale, 2000), vol. 1, 869–72. Angela Aleiss revealed Iron Eyes Cody's identity in the *New Orleans Times-Picayune*, May 26, 1996, and her exposé became the basis for many subsequent reports, especially following the actor's death in 1999. Robert Thompson quoted in Advertising Council, "Public Service Advertising That Changed a Nation," 2004, 7, available online at www.adcouncil.org/files/adweek_report_2004.pdf.

2. On images as moments of visual eloquence, see Robert Hariman and John Louis Lucaites, *No Caption Needed: Iconic Photographs, Public Culture, and Liberal Democracy* (Chicago: University of Chicago Press, 2007). On the Ecological Indian, see also Shepard Krech III, *The Ecological Indian: Myth and History* (New York: W. W. Norton, 1999); and Michael E. Harkin and David Rich Lewis, eds., *Native Americans and the Environment: Perspectives on the Ecological Indian* (Lincoln: University of Nebraska Press, 2007).

3. Tom Lutz, *Crying: The Natural and Cultural History of Tears* (New York: W. W. Norton, 1999), 51–52.

4. Iron Eyes Cody as told to Collin Perry, *Iron Eyes, My Life as a Hollywood Indian* (New York: Everest House, 1982), 269–70.

5. On the Advertising Council, see, for example, Robert Griffith, "The Selling of America: The Advertising Council and American Politics," *Business History Review* 57 (Autumn 1983): 388–412. On Keep America Beautiful, see Heather Rogers, *Gone Tomorrow: The Hidden Life of Garbage* (New York: New Press, 2005), 141–50.

6. David F. Beard, form letter to broadcasters, June 1961; Advertising Council, radio fact sheet for Keep America Beautiful (quotation on 1), both in file 1023, June 1961, 13/2/207, Advertising Council Archives, University Archives, University of Illinois (hereafter ACA).

7. Advertising Council, Radio Fact Sheet for Keep America Beautiful (quotations on 1 and 3), in File 1023, June 1961, 13/2/207, ACA.

8. David F. Beard, form letter to broadcasters, undated, in File 1096, April 1963, 13/2/207, ACA. Examples of advertisements from this campaign are in File 1097, April-June 1963, and File 1107, July-September 1963, 13/2/207, ACA.

9. David F. Beard, form letter to broadcasters, undated, in file 1158, April 1964, 13/2/207, ACA. The second quotation is from a KAB advertisement in file 1164, July-September 1964, 13/2/207, ACA.

10. Beard, form letter to broadcasters, undated, in file 1158, April 1964, 13/2/207, ACA. Examples of the Susan Spotless advertisement are in file 1158, April 1964; file 1164, July–September 1964; file 1166, May 1964; file 1171, June 1964, all in 13/2/207, ACA.

11. KAB Advertisement in file 1446, 3rd quarter 1968, 13/2/207, ACA.
12. Tom Baker, memorandum to National Soft Drink Association members, March 31, 1970 (quotations on 1). I found a copy of this memo in the documents of the FBI Surveillance of Earth Day Activities (4-22-70), which I obtained through a Freedom of Information / Privacy Acts request.
13. Information on the protests in Atlanta and elsewhere contained in the following reports filed by FBI agents: Atlanta (100-8078), April 13, 1970, April 26, 1970, and April 29, 1970; Anchorage (100-2502), April 27, 1970; and Indianapolis (94-425), April 27, 1970. The "trash receptacles" quotation is in the report from Atlanta, April 26. The figure of fifty field offices is taken from R. D. Cotter to C. D. Brennan, memorandum dated April 15, 1971, all in documents of the FBI Surveillance of Earth Day Activities.
14. Tom Baker, memorandum to National Soft Drink Association members, 2.
15. Robin Richman, "Rediscovery of the Redman," *Life*, December 1, 1967, 59. On the counterculture's appropriation of Indianness, see also Philip J. Deloria, *Playing Indian* (New Haven: Yale University Press, 1998), chap. 6.
16. Gary Snyder quoted in "Land Lovers," *Look*, November 4, 1969, 54. The photograph of Snyder appeared on 56–57.
17. The People's Park poster (Frank Bardacke, People's Park Manifesto, 1969) is reproduced in Deloria, *Playing Indian*, 162. Steven V. Roberts, "The Better Earth," *New York Times Magazine*, March 29, 1970, 8.
18. On *Little Big Man*, see, for example, Margo Kasdan and Susan Tavernetti, "Native Americans in a Revisionist Western: *Little Big Man*," in *Hollywood's Indian: The Portrayal of the Native American in Film*, ed. Peter C. Rollins and John E. O'Connor (Lexington: University Press of Kentucky, 1998), 121–36. The quotation is from Stanley Kauffmann's review of the film in the *New Republic*, December 26, 1970, 18.
19. Louis Magnani (vice president of Marsteller) quoted in "Minutes of the Annual Meeting, Board of Directors, The Advertising Council, Inc., March 17, 1971," 7, in box 10, folder Ad Council Minutes, March–April 1971, 13/2/210, ACA.
20. See Edward Buscombe, *"Injuns!" Native Americans in the Movies* (London: Reaktion Books, 2006), 158–64.
21. My interpretation of Iron Eyes Cody as a ghost and time traveler draws on broader studies of Indians in the American imagination, including Philip J. Deloria, *Indians in Unexpected Places* (Lawrence: University Press of Kansas, 2004); and Renée L. Bergland, *The National Uncanny: Indian Ghosts and American Subjects* (Hanover, NH: University Press of New England, 2000).
22. Vine Deloria Jr., "This Country Was a Lot Better Off When the Indians Were Running It," *New York Times Magazine*, March 8, 1970. On Alcatraz and Native American activism during this period, see Troy R. Johnson, *The Occupation of Alcatraz Island: Indian Self-Determination and the Rise of Indian Activism* (Urbana: University of Illinois Press, 1996).

23. Deloria Jr., "This Country."

24. On protests against the film *A Man Called Horse* (1970), see Angela Aleiss, *Making the White Man's Indian: Native Americans and Hollywood Movies* (Westport, CT: Praeger, 2005), 131–33; Cody, *Iron Eyes*, 251.

25. Quotation from NBC News, May 4, 1971, VTNA.

26. "Help Fight Pollution—1971," campaign copy and fact sheet, box 2, folder Campaign Fact Sheet & Copy Guide #49-R-100, 1968–1971, 13/2/210, ACA (first quotation). Advertisement in file 1646, May 1971, 13/2/207, ACA (second quotation). Advertisement reprinted in Peter Harnik, "Debunking Madison Avenue," *Environmental Action*, September 4, 1971, 11 (third quotation).

27. John G. Mitchell, "Keeping America Bottled (and Canned)," *Audubon*, March 1976, 107. Ted Williams, "The Metamorphosis of Keep America Beautiful," *Audubon*, March 1990, 126. For more on the bottle bill debate, see Rogers, *Gone Tomorrow*, 141–51.

28. Harnik, "Debunking Madison Avenue," 11; and Mitchell, "Keeping America Bottled (and Canned)," 106.

29. The first quotation is from Lewis W. Shollenberger to Vincent J. Mullins, July 29, 1976, box 28, folder FCC Petition, 1976, 13/2/305, ACA. The second quotation is from Storyboard for Television Spot, March 1975, box 1, folder Employer Support of the Guard and Reserve, Red Cross, Keep America Beautiful, Help America Work, Rehabilitation of the Handicapped, Forest Fire Prevention, United Negro College Fund, March 1975, 13/2/214, ACA. The final quotation is from a newspaper advertisement in file 1905, April–June 1975, 13/2/207, ACA.

CHAPTER SIX

1. Leo Marx, "American Institutions and Ecological Ideals," *Science*, November 27, 1970, 945–52, reprinted in his *The Pilot and the Passenger: Essays on Literature, Technology, and Culture in the United States* (New York: Oxford University Press, 1988), 139–59 (quotations on 142, 144, 146, 158, and 159); Leo Marx, *The Machine in the Garden: Technology and the Pastoral Ideal in America* (New York: Oxford University Press, 1964).

2. Quotations from Edward D. Berkowitz, *Something Happened: A Political and Cultural Overview of the Seventies* (New York: Columbia University Press, 2006), 10, 84, and 87. For similar comments about the significance of 1974 as a dividing line as well as the increasing prominence of the rhetoric of individual responsibility, see Philip Jenkins, *Decade of Nightmares: The End of the Sixties and the Making of Eighties America* (New York: Oxford University Press, 2006), especially 5 and 149.

3. My critique of environmentalism and environmental policy is informed by Robert Gottlieb, *Forcing the Spring: The Transformation of the American Environmental Movement* (Washington: Island Press, 1993).

4. First quotation from J. C. Dyer, "The History of the Recycling Symbol: How Gary Anderson Designed the Recycling Symbol," http://www.dyer -consequences.com/recycling_symbol.html; second quotation from Penny Jones and Jerry Powell, "Gary Anderson Has Been Found!" *Resource Recycling*, May 1999, 25.

5. Ivars Peterson, *Fragments of Infinity: A Kaleidoscope of Math and Art* (New York: John Wiley & Sons, 2001), 136 and 144. On Escher's interest in the infinite and in the Möbius strip, see also Eli Maor, *To Infinity and Beyond: A Cultural History of the Infinite* (Princeton, NJ: Princeton University Press, 1991), 141–42 and 164–78.

6. Marianne Hancock, review of *The World of M. C. Escher*, ed. J. L. Locher, *Arts Magazine*, November 1972, 91; Thomas Albright, "Visuals," *Rolling Stone*, February 21, 1970, 40–41 (quotation on 40). The Escher images appeared in *Saturday Review*, March 7, 1970, for a special issue titled "Cleaning Humanity's Nest," and in *Ramparts*, November 1969, for a special issue reprinted as *Eco-Catastrophe*, ed. editors of *Ramparts* (New York: Harper and Row, 1970). Advertisements for *The Graphic Work of M. C. Escher* appeared in *Whole Earth Catalog*, Spring 1969 and Fall 1969, 65.

7. Anderson quoted in Jones and Powell, "Gary Anderson Has Been Found!" 26. J. L. Locher, "The Work of M. C. Escher," in *The World of M. C. Escher*, ed. J. L. Locher (New York: Harry N. Abrams, 1974), 5–14 (quotation on 8); and Peterson, *Fragments of Infinity*, 141.

8. On the concept of virtuous consumption, see Marguerite S. Shaffer, *See America First: Tourism and National Identity, 1880–1940* (Washington: Smithsonian Institution Press, 2001), 5.

9. The first quotation is from Reyner Banham, introduction to *The Aspen Papers: Twenty Years of Design Theory from the International Design Conference in Aspen*, ed. Reyner Banham (New York: Praeger, 1974), 14; Alice Twemlow, "I Can't Talk to You If You Say That: An Ideological Collision at the International Design Conference at Aspen, 1970," *Design and Culture* 1 (March 2009): 23–49 (quotations on 24 and 25).

10. Humphrey quoted in Twemlow, 34.

11. Baudrillard quoted in "The Environmental Witch-Hunt: Statement by the French Group" (1970), reprinted in *The Aspen Papers*, 208–10.

12. "Happy Birthday Earth Day!" *Look*, May 4, 1971, 19. The quotations concerning Modesto, California, are from the following article in the same issue: Margaret M. McGlynn, "Garbage Power!" 26, 29, 30. The "cleanup crusade" quotation is from a caption in the same issue, 2. For similar coverage in *Life* magazine during 1971, see "More Oil for Our Troubled Waters," *Life,* February 5, 1971, 36–43; and "A Clean-Up Mood Sweeps the Nation," *Life*, March 5, 1971, 30–35. See also an advertisement for *Seventeen* magazine, entitled "What's Right with America," in *Advertising Age*, June 21, 1971, 9. On Cliff Humphrey's successful efforts to start curbside recycling programs in Modesto and elsewhere, see Larry Borowsky, "The

Consummate Recycler," *Environmental Action*, March/April 1989, reprinted in *Learning to Listen to the Land*, ed. Bill Willers (Washington: Island Press, 1991), 248–54.

13. Humphrey quoted in McGlynn, "Garbage Power!" 26. On the counterculture, the *Whole Earth Catalog*, and environmental hope, see also Andrew G. Kirk, *Counterculture Green: The Whole Earth Catalog and American Environmentalism* (Lawrence: University Press of Kansas, 2007).

14. Ted Steinberg, "Can Capitalism Save the Planet? On the Origins of Green Liberalism," *Radical History Review* 107 (Spring 2010): 7–24 (quotation on 13). The second quotation (of the environmental activist Patricia Taylor) appears in Heather Rogers, *Gone Tomorrow: The Hidden Life of Garbage* (New York: The New Press, 2005), 172. On the limits of recycling, see also Timothy W. Luke, "Green Consumerism: Ecology and the Ruse of Recycling," in *In the Nature of Things: Language, Politics, and the Environment*, ed. Jane Bennett and William Chaloupka (Minneapolis: University of Minnesota Press, 1993), 154–72; and Stephen Horton, "Rethinking Recycling: The Politics of the Waste Crisis," *Capitalism Nature Socialism* 6 (March 1995): 1–19.

15. On the significance of "pollution" in media coverage and popular views of the environmental crisis during this period, see also Christopher C. Sellers, *Crabgrass Crucible: Suburban Nature and the Rise of Environmentalism in Twentieth-Century America* (Chapel Hill: University of North Carolina Press, 2012), 254–65.

16. Quotation from "More Oil for Our Troubled Waters," 38–39. On media coverage of volunteers trying to rescue oil-soaked birds and marine mammals, see also Kathryn Morse, "There Will Be Birds: Images of Oil Disasters in the Nineteenth and Twentieth Centuries," *Journal of American History* 99 (June 2012): 124–34, esp. 124 and 131–33. On the "Hard Hat Riots," see, for example, Jefferson Cowie, "Nixon's Class Struggle: Romancing the New Right Worker, 1969–1973," *Labor History* 43 (August 2002): 257–83.

17. See also Paul Sabin, "Crisis and Continuity in U.S. Oil Politics, 1965–1980," *Journal of American History* 99 (June 2012): 177–86, esp. 182.

CHAPTER SEVEN

1. For examples of gas line imagery, see some of the illustrations that accompany the following magazine articles: "Long Lines, Short Tempers," *Newsweek*, March 4, 1974, 65; "The Times They Are A-Changin,'" *Newsweek*, February 18, 1974, 19; "The Whirlwind Confronts the Skeptics," *Time*, January 21, 1974, 22–23; and "A Global Deal on Prices?," *Time*, January 14, 1974, 15. For examples from 1979, see "Gas: A Long Dry Summer?," *Time*, May 21, 1979, 15 and 16; and "The Great Energy Mess," *Time*, July 2, 1979, 15 and 20.

2. On popular framings of the energy crisis as solely a question of supply, see also David E. Nye, *Consuming Power: A Social History of American Energies* (Cambridge, MA: MIT Press, 1998), chap. 8. On automobility, identity, and citizenship, see Cotten Seiler, *Republic of Drivers: A Cultural History of Automobility in America* (Chicago: University of Chicago Press, 2008). On the energy crisis and the discourse of individual blame and responsibility, see also Natasha Zaretsky, *No Direction Home: The American Family and the Fear of National Decline, 1968–1980* (Chapel Hill: University of North Carolina Press, 2007), chap. 2. On the 1973–74 oil shock, including the meanings of gas lines, see also Mark Fiege, *The Republic of Nature: An Environmental History of the United States* (Seattle: University of Washington Press, 2012), chap. 9.

3. *Network*, DVD, directed by Sidney Lumet (1976; Burbank, CA: Warner Home Video and Turner Entertainment, 2006). Mark Shiel, "A Nostalgia for Modernity: New York, Los Angeles, and American Cinema in the 1970s," in *Screening the City*, ed. Mark Shiel and Tony Fitzmaurice (New York: Verso, 2003), 160–79; quotation on 173.

4. Advertising Council, story board titled "George C. Scott: 'Empty Schoolroom,'" box 1, folder Energy Conservation, February 1974, 13/2/214, ACA; Advertising Council, "How Do You Urge Americans to Save Energy? Advertising Council Volunteer Agency Cunningham & Walsh Meets the Challenge," box 1, folder Energy Conservation, 1974–1980, 13/2/279, ACA; Advertising Council, "6 ways not to be fuelish," file 1809, 13/2/207, ACA.

5. Richard Nixon, "Address to the Nation about Policies to Deal with the Energy Shortages," November 7, 1973, in John T. Woolley and Gerhard Peters, *The American Presidency Project*, Santa Barbara, CA, available online at http://www.presidency.ucsb.edu/ws/?pid=4034.

6. Gerald Ford, "Address before a Joint Session of the Congress Reporting on the State of the Union," January 15, 1975, in John T. Woolley and Gerhard Peters, *The American Presidency Project*, Santa Barbara, CA, available online at http://www.presidency.ucsb.edu/ws/?pid=4938; Cunningham & Walsh Inc., press release titled "New FEA TV Commercial Features Boy's Plea to Conserve Energy," in box 14, folder Energy (Federal Energy Administration & Cunningham & Walsh), April/May 1975, 13/2/220, ACA; Les Gapay, "FEA Says Campaign to Save Energy Led to Waste of Money," *Wall Street Journal*, October 26, 1976 ("historic sites" quotation).

7. Peter Harnik, "Jack Anderson Script," August 5, 1976, as delivered to Ad Council's Washington office by Peter Harnik, copy in box 29, folder Keep America Beautiful, 1976, 13/2/305, ACA. See also Peter Harnik, "The Junking of an Anti-Litter Lobby," *Business and Society Review*, Spring 1977, 47–51.

8. For a brief history of nuclear energy, including the commercial nuclear power industry, see J. Samuel Walker, "From the 'Atomic Age' to the

'Anti-Nuclear Age': Nuclear Energy in Politics, Diplomacy, and Culture," in *A Companion to Post-1945 America*, ed. Jean-Christophe Agnew and Roy Rosenzweig (Malden, MA: Blackwell Publishers, 2002), 501–18 (Strauss is quoted on 514). On the ecological significance of the names of local anti-reactor groups, see also Lisa Lynch, "'We Don't Wanna Be Radiated': Documentary Film and the Evolving Rhetoric of Nuclear Energy Activism," *American Literature* 84 (June 2012): 327–51, esp. 335 and 331. The poster quotation is from John Wills, *Conservation Fallout: Nuclear Protest at Diablo Canyon* (Reno: University of Nevada Press, 2006), 83.

9.   Amory B. Lovins, "Energy Strategy: The Road Not Taken?" *Foreign Affairs* 55 (October 1976): 65–96 (quotations on 65, 66, and 74–75).

10. Ava Swartz, "Ads for Outgroups," *Columbia Journalism Review* 12 (March/ April 1974): 12–15 ("It's not your fault" quotations on 15). David L. Paletz, Roberta E. Pearson, and Donald L. Willis, *Politics in Public Service Advertising on Television* (New York: Praeger Publishers, 1977), 109 (nuclear power quotation).

11. Thomas Asher, "Smoking Out Smokey the Bear," *MORE*, March 1972, 12–14 (first quotation on 13); William D. Lutz, "'The American Economic System': The Gospel According to the Advertising Council," *College English* 38 (April 1977): 860–65 (second quotation on 861).

12. Public Media Center, et al., "Petition to Institute a Notice of Inquiry and Proposed Rulemaking on the Airing of Public Service Announcements by Broadcast Licensees," presented to the Federal Communications Commission, 1976, 1, copy in box 28, folder FCC Media Access Project, 1976, 13/2/305, ACA. The list of petitioners appears at the beginning of the petition and is not paginated.

13. Public Media Center, et al., "Petition," 27–28.

14. Public Media Center, et al., "Petition," 8.

15. Lewis W. Shollenberger to Vincent J. Mullins, July 29, 1976, box 28, folder FCC Petition, 1976, 13/2/305, ACA.

16. Public Media Center, et al., "Petition," 1.

CHAPTER EIGHT

1.   The first quotation is from "Columbia Pulling Out Stops to Hype 'China Syndrome,'" *Variety*, February 23, 1979, copy in *The China Syndrome* production file, Margaret Herrick Library, Academy of Motion Picture Arts and Sciences, Beverly Hills, California (hereafter CSPF); the other quotations are from Aljean Harmetz, "First Tease-by-TV Ad Campaign?" *New York Times*, March 8, 1979. One version of the trailer can be seen on *The China Syndrome: Creating a Controversy*, directed by Laurent Bouzereau, a special feature on *The China Syndrome*, DVD, directed by James Bridges (1979; Culver City, CA: Columbia Pictures/Columbia TriStar Home Entertainment, 2004).

2. Most quotations in this paragraph are from William K. Knoedelseder Jr. and Ellen Farley, "'China Syndrome' Stirs Wave of Pro-Nuclear Protests," *Los Angeles Times*, March 25, 1979, CSPF. The quotations from the energy trade magazine appear in Aljean Harmetz, "Fallout from 'China Syndrome' Has Already Begun," *New York Times*, March 11, 1979.

3. Vincent Canby, "Film: Nuclear Plant Is Villain in 'China Syndrome,'" *New York Times*, March 16, 1979; Richard Schickel, "Art: An Atom-Powered Thriller," *Time*, March 26, 1979, 54.

4. David Burnham, "Nuclear Experts Debate 'The China Syndrome,'" *New York Times*, March 18, 1979.

5. Ralph Kaminsky, "Col Halts Promo Push for 'Syndrome' After Pennsylvania Nuclear Accident," *Box Office*, April 23, 1979, CSPF; Aljean Harmetz, "When Nuclear Crisis Imitates a Film," *New York Times*, April 4, 1979; "Pa. Crisis a Powerful Trailer for 'China Syndrome,'" *Variety*, April 4, 1979, CSPF. For other scholarly sources that examine *The China Syndrome*, see David Ingram, *Green Screen: Environmentalism and Hollywood Cinema* (Exeter, UK: University of Exeter Press, 2000), 168–73; John Wills, "Celluloid Chain Reactions: *The China Syndrome* and Three Mile Island," *European Journal of American Culture* 25 (August 2006): 109–22; and Marsha Weisiger, "When Life Imitates Art," *Environmental History* 12 (April 2007): 383–85.

6. Quotations from Aaron Latham, "Hollywood vs. Harrisburg," *Esquire*, May 22, 1979, 78.

7. On the Brookhaven Report, see McKinley C. Olson, "The Hot River Valley," originally published in *The Nation*, August 3, 1974, 69–85, reprinted in *America's Energy: Reports from* The Nation *on 100 Years of Struggles for the Democratic Control of Our Resources*, ed. Robert Engler (New York: Pantheon, 1980), 321–35, esp. 328–30; Union of Concerned Scientists, *The Risks of Nuclear Power Reactors: A Review of the NRC Reactor Safety Study, WASH-1400 (NUREG-75/014)* (Cambridge, MA: Union of Concerned Scientists, 1977), 3 (quotation). After gaining access to the Brookhaven report, the Union of Concerned Scientists, along with Ralph Nader, one of its leading spokespersons, began using this line about Pennsylvania. In addition to *The Risks of Nuclear Power Reactors*, see Daniel F. Ford and Henry W. Kendall to Robert Byrd, October 3, 1973, in box 32, folder "B" Correspondence, 1973–1976; and Ralph Nader, letter to the residents of South Shore, July 22, 1974, in box 37, folder Nader at Plymouth, both in the papers of the Union of Concerned Scientists (MC 434), Institute Archives and Special Collections, MIT Libraries, Massachusetts Institute of Technology.

8. Gray quoted in Knoedelseder Jr. and Farley, "'China Syndrome' Stirs Wave."

9. Douglas quoted in Ben Fong-Torres, "'The China Syndrome,'" *Rolling Stone*, April 5, 1979, 50–55; quotations on 51–52.

10. Gilbert quotations from Barbara Zheutlin and Jim Richardson, "Hollywood's Progressive Producer: An Interview with Bruce Gilbert," *Cineaste*, Fall 1979, 2–9; quotations on 2. Thomas Kiernan, *Jane Fonda: Heroine for Our Time* (New York: Delilah Books, 1982), 301. On Fonda and politics, see also Steven J. Ross, *Hollywood Left and Right: How Movie Stars Shaped American Politics* (New York: Oxford University Press, 2011), chap. 6.

11. On Widener and *Powers That Be*, see James Real, "#1 on the Nuclear Blacklist: The Purge of Filmmaker Don Widener," *Mother Jones*, December 1978, 20–32. Widener quoted in Eleanor Smith, "If At First You Don't Succeed, Play Dirty," originally published in *Not Man Apart*, September/October 1978, 4–5, 10, 19, reprinted in *Atom's Eve: Ending the Nuclear Age: An Anthology*, ed. Mark Reader with Ronald A. Hardert and Gerald L. Moulton (New York: McGraw-Hill, 1980), 153–61; quotation on 153.

12. Quotation from Fong-Torres, 52.

13. David A. Cook, *Lost Illusions: America Cinema in the Shadow of Watergate and Vietnam, 1970–1979* (Berkeley: University of California Press, 2000), 197.

14. James Bridges, "Eyewitness," script, first draft with revisions, 1977, 5a, in box 22B, folder 17, Michael Douglas papers, 1934–1980, Wisconsin Center for Film and Theater Research, Madison, WI (hereafter MDP).

15. Fonda quoted in *The China Syndrome: A Fusion of Talent*, directed by Laurent Bouzereau, a special feature on *The China Syndrome*, DVD, 2004.

16. Gary Crowdus, review of *The China Syndrome*, *Cineaste*, Spring 1979, 45–47; quotation on 46.

17. Gregory C. Minor to Harry H. Hendon, February 2, 1976, letter of resignation reprinted as appendix to pamphlet titled *Testimony of Dale G. Bridenbaugh, Richard B. Hubbard, Gregory C. Minor Before the Joint Committee on Atomic Energy, February 18, 1976* (Cambridge, MA: Union of Concerned Scientists, 1976), copy in box 22B, folder 19, MDP. On the GE engineers' involvement with the film, see also Ingram, *Green Screen*, 170 and 173.

18. George F. Will, "A Film about Greed," *Newsweek*, April 2, 1979, 96; Patricia A. Corrigan, letter to editor, *Newsweek*, April 23, 1979, 9; Arnold Ahlert, letter to editor, *Newsweek*, May 7, 1979, 9; and Robert E. Rodi, letter to editor, *Newsweek*, April 23, 1979, 4 and 9; quotation on 9.

19. Robert Hatch, "Films," *The Nation*, March 31, 1979, 347–50; quotations on 347 and 348.

20. Murray Hausknecht, "Meltdowns and Ideologies," *Dissent*, Fall 1979, 468–71; quotations on 468 and 469.

21. Hatch, 348 and 350; Hausknecht, 470 and 471.

22. Mike and Carol Gray, "The China Syndrome," script, third draft, October 26, 1975, 60, copy in box 22B, folder 15, MDP; for mention of solar energy see, for example, Bridges (same script as in note 13), 86.

23. James Bridges, "The China Syndrome," as-shot screenplay, May 1978, 73 and 71, copy in James Bridges papers, American Heritage Center, University of Wyoming.

24. Crowdus, 46; John Mariani, "Interview: James Bridges on Directing *The China Syndrome*," *Millimeter*, July 1979, page number illegible in copy, CSPF.

25. Wayne J. McMullen, "*The China Syndrome*: Corruption to the Core," *Literature/Film Quarterly* 23 (January 1995): 55–62 (quotation on 59); and Doug Zwick, "*The China Syndrome*: The Genre Syndrome," *Jump Cut* 22 (May 1980): 5–6.

26. On Mothers for Peace, see John Wills, *Conservation Fallout: Nuclear Protest at Diablo Canyon* (Reno: University of Nevada Press, 2006), chap. 3.

27. Fonda made this argument in an interview included on *The China Syndrome: Creating a Controversy*, directed by Laurent Bouzereau.

28. Gray, 36–37.

29. Jane M. Gaines, "Political Mimesis," in *Collecting Visible Evidence*, ed. Jane M. Gaines and Michael Renov (Minneapolis: University of Minnesota Press, 1999), 84–102, esp. 89–93 (quotation on 89). My discussion of antinuclear documentaries draws on the perceptive analysis of Lisa Lynch, "'We Don't Wanna Be Radiated': Documentary Film and the Evolving Rhetoric of Nuclear Energy Activism," *American Literature* 84 (June 2012): 327–51 (quotation from film on 333).

30. First quotation from *WIN* staff, "An Editorial," *WIN*, April 12, 1979, 4–5 (quotation on 4); second and third quotations from Linda Blackaby, Dan Georgakas, and Barbara Margolis, *In Focus: A Guide to Using Films* (New York: New York Zoetrope, 1980), 63. For other examples of protestors leafleting theaters, see http://archives.library.wisc.edu/uw-archives/exhibits/protests/1970s.html (photograph of students in Madison, WI, March 13, 1979); and http://vimeo.com/8243629 (video from Swindon, UK). On the Abalone Alliance protest in San Luis Obispo, see Barbara Epstein, *Political Protest and Cultural Revolution: Nonviolent Direct Action in the 1970s and 1980s* (Berkeley: University of California Press, 1991), 100; Robert Gottlieb, *Forcing the Spring: The Transformation of the American Environmental Movement* (Washington: Island Press, 1993), 181; and Wills, *Conservation Fallout*, 93.

CHAPTER NINE

1. CBS News, March 28, 1979, available from VTNA. For a history of the accident, see J. Samuel Walker, *Three Mile Island: A Nuclear Crisis in Historical Perspective* (Berkeley: University of California Press, 2004).

2. ABC News, March 28, 1979 (both quotations); NBC News, March 28, 1979; both from VTNA. On television news coverage of Three Mile Island, see also Dan Nimmo and James E. Combs, *Nightly Horrors: Crisis Coverage by Television Network News* (Knoxville: University of Tennessee Press, 1985), chap. 2.

3. Mary Ann Doane, "Information, Crisis, Catastrophe," in *Logics of Television: Essays in Cultural Criticism,* ed. Patricia Mellencamp (Bloomington: Indiana University Press, 1990), 222–39 (quotations on 222, 223, and 230).

4. CBS News, March 30, 1979, from VTNA.

5. CBS News, March 30, 1979; NBC News, March 30, 1979, ABC News, March 30, 1979; all from VTNA.

6. A. Kent MacDougall, "Media Cast Largely Uncritical Eye on Nuclear Power Industry Until . . . ," *Los Angeles Times,* February 6, 1980 (*20/20* quotation); and "Beyond 'The China Syndrome,'" *Newsweek,* April 16, 1979, 31 (all other quotations).

7. Nimmo and Combs, 69, 70, and 71.

8. *Newsweek,* April 9, 1979, cover ("Nuclear Accident"); *Time,* April 9, 1979, cover ("Nuclear Nightmare"). See also *Life,* May 1979, cover. Gregory Heisler, whose Three Mile Island photograph appeared on the cover of *Life,* emphasized the significance of the red lights. "I knew the red lights blinking on the towers would set off the red *Life* logo. I was shooting for a cover here." See Laurence Shames, "Professionally Tailored," *American Photographer,* August 1986, 36–48 (quotation on 44).

9. Walker, 10.

10. Quotation from Robert Hariman and John Louis Lucaites, *No Caption Needed: Iconic Photographs, Public Culture, and Liberal Democracy* (Chicago: University of Chicago Press, 2007), 277–78. In this passage, Hariman and Lucaites are discussing not Three Mile Island in particular, but rather media coverage of technological risk more generally.

11. As Hariman and Lucaites argue in *No Caption Needed,* the iconicity of an image can be demonstrated in part through its circulation in other visual media, including cartoons.

12. *Mad,* October 1979, back cover.

13. On their tour of the Trojan Nuclear Power plant, see, for example, Aaron Latham, "Hollywood vs. Harrisburg," *Esquire,* May 22, 1979, 80; Michael Douglas also discusses their visit in an interview included on *The China Syndrome: Creating a Controversy,* directed by Laurent Bouzereau, a special feature on *The China Syndrome,* DVD, directed by James Bridges (1979; Culver City, CA: Columbia Pictures / Columbia TriStar Home Entertainment, 2004). On Trojan's cooling towers, which were imploded in 2006, see Jonathan Martin, "Vestige of Region's Nuclear Designs to be Imploded Sunday Morning," *Seattle Times,* May 19, 2006.

14. On the movement against nuclear power, see Barbara Epstein, *Political Protest and Cultural Revolution: Nonviolent Direct Action in the 1970s and 1980s* (Berkeley: University of California Press, 1991), chaps. 2 and 3; Thomas Raymond Wellock, *Critical Masses: Opposition to Nuclear Power in California, 1958–1978* (Madison: University of Wisconsin Press, 1998); John Wills, *Conservation Fallout: Nuclear Protest at Diablo Canyon* (Reno: University of

Nevada Press, 2006); and Robert Surbrug Jr., *Beyond Vietnam: The Politics of Protest in Massachusetts, 1974–1990* (Amherst: University of Massachusetts Press, 2009).

15. ABC News, May 6, 1979, VTNA; "'Hell No, We Won't Glow,'" *Time*, May 21, 1979, 17–18 (quotations on 18); "Fallout from the Nuke-in," *Newsweek*, May 21, 1979, 34–35 (quotation on 34).

16. "'Hell No, We Won't Glow'" ("60s spirit" and "nostalgic search" quotations); ABC News, May 6, 1979; NBC News, May 6, 1979, both from VTNA; Wendell Rawls Jr., "65,000 Demonstrate at Capitol to Halt Atomic Power Plants," *New York Times*, May 7, 1979 ("graduates" quotation).

17. CBS News, May 6, 1979, VTNA ("mostly middle-class" quotation); Rawls Jr., "65,000 Demonstrate" ("white" quotation); "'Hell No, We Won't Glow'" ("noticeably lacking" quotation).

18. Valerie L. Kuletz, *The Tainted Desert: Environmental and Social Ruin in the American West* (New York: Routledge, 1998), 26–27. See also Marjane Ambler, *Breaking the Iron Bonds: Indian Control of Energy Development* (Lawrence: University Press of Kansas, 1990), 174–76; and Wm. Paul Robinson, "Uranium Production and Its Effects on Navajo Communities along the Rio Puerco in Western New Mexico," in *Race and the Incidence of Environmental Hazards: A Time for Discourse*, ed. Bunyan Bryant and Paul Mohai (Boulder, CO: Westview Press, 1992), 153–62.

19. ABC News, August 27, 1979, VTNA.

20. On nuclear colonialism, see, for example, Danielle Endres, "The Rhetoric of Nuclear Colonialism: Rhetorical Exclusion of American Indian Arguments in the Yucca Mountain Nuclear Waste Siting Decision," *Communication and Critical/Cultural Studies* 6 (March 2009): 39–60.

21. Michael Aaron Weiss and Stanley David Weiss, letter to editor, *Commentary*, September 1979, 4 and 8 (quotation on 8).

22. First and second quotations from Walker, 224; third quotation from Christian Joppke, *Mobilizing Against Nuclear Energy: A Comparison of Germany and the United States* (Berkeley: University of California Press, 1993), 144.

23. Fonda quotation from ABC News, September 23, 1979, VTNA; the "No Nukes" and "Turn On the Sun" banners appeared on NBC News, September 23, 1979, VTNA; the Frisbee appeared on CBS News, April 7, 1979, VTNA.

CHAPTER TEN

1. Denis Hayes, form letter about Sun Day, n.d., copy in box 33, folder Alliance to Conserve Energy, January–June 1978, 13/2/305, ACA.

2. Harnik quotation in Gladwin Hill, "'Earth Day' Backer Planning 'Sun Day,'" *New York Times*, October 16, 1977; Hayes quotation in "All Hail The Sun!" *Newsweek*, May 15, 1978, 30–31 (quotation on 30).

3. Peter Harnik, "Here Comes The Solar Age," *Sun Day Times*, February 1978, 1 and 4 (quotation on 4), copy in box DB 18, folder 12, Dorothy Boberg Collection, 1957–1986, Special Collections and Archives, Oviatt Library, California State University, Northridge. Marlene Cimons, "The Goal of Sun Day: Solar Power," *Los Angeles Times*, December 1, 1977 (second quotation).

4. Deborah Baldwin, "Sun Day is Wednesday, May 3," *Environmental Action*, October 22, 1977, 14–15 ("apple pie" and "noncontroversial" quotes on 14). The other quotations are from Mina Hamilton, "Sun Day: Dawn of an Era or Just Another Wednesday?," *Not Man Apart*, mid-August/September 1978, 10 and 11. On the techno-determinism of some soft-path advocates, see also Christian Joppke, *Mobilizing against Nuclear Energy: A Comparison of Germany and the United States* (Berkeley: University of California Press, 1993), 74 and 77.

5. Hayes quoted on CBS News, May 3, 1978, from VTNA; Harnik, "Here Comes The Solar Age," 1 (second quotation).

6. ABC News, April 12, 1973, VTNA. For more on Thomason's career, see Daniel Behrman, *Solar Energy: The Awakening Science* (1976; London: Routledge & Kegan Paul, 1979), 113–29.

7. ABC News, March 30, 1978, VTNA. For other examples of TV coverage that applied the individualist frame, see CBS News, May 27, 1975; ABC News, February 4, 1977; and NBC News, March 3, 1977, all from VTNA.

8. Peter Calthorpe and Susan Benson, "Beyond Solar Suburbia," originally published in *Progressive Architecture*, April 1979, reprinted in *RAIN* August/September 1979, 12–14 (quotation on 12).

9. Diane Schatz's posters can be found in multiple issues of *RAIN*, including on the covers and in the interior pages of the April 1976 and October 1978 issues. *RAIN* also sold the posters and reprinted them in other publications, including *RAINBOOK: Resources for Appropriate Technology*, ed. editors of *RAIN* and Lane deMoll (New York: Schocken Books, 1977). I first saw Schatz's poster in Andrew Kirk, "'Machines of Loving Grace': Alternative Technology, Environment, and the Counterculture," in *Imagine Nation: The American Counterculture of the 1960s and '70s*, ed. Peter Braunstein and Michael William Doyle (New York: Routledge, 2002), 353–78.

10. Denis Hayes, *Rays of Hope: The Transition to a Post-Petroleum World* (New York: W. W. Norton, 1977), 77–79 and 87.

11. Hayes, *Rays of Hope*, 207.

12. Norma Skurka and Jon Naar, *Design for a Limited Planet: Living with Natural Energy* (New York: McGraw-Hill Book Company, 1976), 192–97.

13. The members of Sun Day's board of directors are listed on the form letter about Sun Day.

14. Harvey Wasserman, "Unionizing Ecotopia," *Mother Jones*, June 1978, 30–37 (first quotation on 32). Douglas Fraser, "The Sun Offers Energy and

Jobs," in *Accidents Will Happen: The Case against Nuclear Power,* ed. Lee Stephenson and George R. Zachar (New York: Harper & Row, 1979), 239–47; quotation on 239. See also Douglas Fraser, "Solar Means More Jobs," *Sun Day Times,* March 1978, 1 and 3, copy in Series 1: Organizational Materials, "Sun Day," Alternative Energy Coalition Records, 1960–81, MS 586, Special Collections and University Archives, University of Massachusetts Amherst Libraries. On Environmentalists for Full Employment, see also Robert Gottlieb, *Forcing the Spring: The Transformation of the American Environmental Movement* (Washington: Island Press, 1993), 290–91; and Robert Surbrug Jr., *Beyond Vietnam: The Politics of Protest in Massachusetts, 1974–1990* (Amherst: University of Massachusetts Press, 2009), 49 and 80.

15. Jefferson Cowie, *Stayin' Alive: The 1970s and the Last Days of the Working Class* (New York: The New Press, 2010), chap. 6; quotations on 262. Fraser is quoted in Cowie, 299.

16. Luther J. Carter, "Sun Day Seen as More Potent Politically than Earth Day," *Science,* April 14, 1978, 185–87; quotations on 185 and 186.

17. ABC News, CBS News, NBC News, all from May 3, 1978, VTNA. The sundial quotation appeared in "Having Fun with the Sun," *Time,* May 15, 1978, 19.

18. Hamilton, "Sun Day," 10.

19. ABC News, May 3, 1978, VTNA.

20. NBC News, May 3, 1978, VTNA.

21. "All Hail The Sun!" 30–31.

22. Cato, "Letter from Washington," *National Review,* May 26, 1978, 632.

23. Langdon Winner, "Building the Better Mousetrap," originally published in 1980, reprinted in *The Whale and the Reactor: A Search for Limits in an Age of High Technology* (Chicago: University of Chicago Press, 1986), 61–84; quotations on 66 and 79. The "ecological individualism" and "technological optimism" quotations are from Andrew G. Kirk, *Counterculture Green: The Whole Earth Catalog and American Environmentalism* (Lawrence: University Press of Kansas, 2007), 52 and 55.

24. On the MUSE concerts, see *No Nukes: The MUSE Concert,* VHS, directed by Danny Goldberg, Julian Schlossberg, and Anthony Potenza (1980; Farmington Hills, MI: CBS/FOX Video Music, 1985). The "copper-colored art-deco sun" quote is from Stephen Davis, "Blues for MUSE," *New Age,* April 1980, 54.

25. NBC News, April 5, 1979, VTNA.

26. On Carter's announcement of the plan to install solar panels, see "All Hail the Sun!" 31; Martin Tolchin, "Carter Orders a Rise for Solar Research," *New York Times,* May 4, 1978; and "A Presidential Bow toward the Sun," *New York Times,* May 7, 1978. Carter had earlier "rejected as too costly a solar water-heating system for the White House." See "Now, It's Sun Day," *Newsweek,* April 24, 1978, 38.

CHAPTER ELEVEN

1. Jimmy Carter, "Solar Energy Remarks Announcing Administration Proposals," June 20, 1979, in John T. Woolley and Gerhard Peters, *The American Presidency Project* (online), Santa Barbara, CA, available online at http://www.presidency.ucsb.edu/ws/?pid=32500. On the dedication ceremony, see also Eliot Marshall, "Carter Hails Solar Age in Presolstice Rite," *Science*, July 6, 1979, 21–24.

2. Ted Shelton, "Greening the White House: Executive Mansion as Symbol of Sustainability," *Journal of Architectural Education* 60 (May 2007): 31–38, esp. 32.

3. *A Road Not Taken*, directed by Christina Hemauer and Roman Keller, 2010; Bill McKibben, "Bring Solar Power Back to the White House," *Washington Post*, September 10, 2010. On the 2013 installation of solar panels, see Juliet Eilperin, "White House Solar Panels Being Installed This Week," *Washington Post*, August 15, 2013.

4. See, for example, the negative comments made by solar power advocates who attended the dedication ceremony in Marshall, "Carter Hails Solar Age," 21. Quotation from Samuel P. Hays, *Beauty, Health, and Permanence: Environmental Politics in the United States, 1955–1985* (New York: Cambridge University Press, 1987), 241–42.

5. On Carter's rhetoric, see Daniel T. Rodgers, *Age of Fracture* (Cambridge, MA: Harvard University Press, 2011), chap. 1. Rodgers also compares Carter to Mister Rogers on 34. On Carter's energy policies, see also John C. Barrow, "An Age of Limits: Jimmy Carter and the Quest for a National Energy Policy," in *The Carter Presidency: Policy Choices in the Post-New Deal Era*, ed. Gary M. Fink and Hugh Davis Graham (Lawrence: University Press of Kansas, 1998), 158–78.

6. CBS News Special Report, "Carter Fireside Address," February 2, 1977, available from VTNA.

7. First and second quotations from Jimmy Carter, "Report to the American People: Remarks from the White House Library," February 2, 1977, in Woolley and Peters, *The American Presidency Project*, available online at http://www.presidency.ucsb.edu/ws/?pid=7455; third and fourth quotations from Jimmy Carter, "The Energy Problem: Address to the Nation," April 18, 1977, in Woolley and Peters, *The American Presidency Project*, available online at http://www.presidency.ucsb.edu/ws/?pid=7369.

8. Janet Marinelli, "Carter's Phony Fuels," *Environmental Action*, September 1979, 25–26; quotation on 25. As the historian Daniel Horowitz notes, the speech "was known by various names, including 'Energy and the Crisis of Confidence' and the 'malaise speech.'" See *Jimmy Carter and the Energy Crisis of the 1970s: The "Crisis of Confidence" Speech of July 15, 1979: A Brief History with Documents*, ed. Daniel Horowitz (Boston: Bedford / St. Mar-

tin's, 2005), 108. In the text of this chapter I follow Horowitz in referring to it as "The Crisis of Confidence." However, the citations for the speech (beginning in note 12) will refer to the title used at *The American Presidency Project* ("Address to the Nation on Energy and National Goals: The 'Malaise' Speech'").

9. Sean Wilentz, *The Age of Reagan: A History, 1974–2008* (New York: Harper-Collins, 2008), 98.

10. Kevin Mattson, *"What the Heck Are You Up To, Mr. President?" Jimmy Carter, America's "Malaise," and the Speech That Should Have Changed the Country* (New York: Bloomsbury, 2009), 8.

11. For an important exception to the overall pattern of historians neglecting the environmentalist and leftist critiques of Carter, see some of the sources collected in *Jimmy Carter and the Energy Crisis of the 1970s*, ed. Daniel Horowitz, esp. 129–33 and 138–42.

12. For examples of gas line imagery, see "Gas: A Long Dry Summer?," *Time*, May 21, 1979, 15 and 16; and "The Great Energy Mess," *Time*, July 2, 1979, 15 and 20. Jimmy Carter, "Address to the Nation on Energy and National Goals: 'The Malaise Speech,'" July 15, 1979, in Woolley and Peters, *The American Presidency Project*, available online at http://www.presidency.ucsb .edu/ws/?pid=32596.

13. Christopher Lasch, *The Culture of Narcissism: American Life in an Age of Diminishing Expectations* (New York: W. W. Norton, 1979); on Lasch's book and Carter's speech, see also Daniel Horowitz, *The Anxieties of Affluence: Critiques of American Consumer Culture, 1939–1979* (Amherst: University of Massachusetts Press, 2004), chaps. 7 and 8; and Natasha Zaretsky, *No Direction Home: The American Family and the Fear of National Decline, 1968–1980* (Chapel Hill: University of North Carolina Press, 2007), chap. 5; Carter, "Address to the Nation on Energy and National Goals."

14. Carter, "Address to the Nation on Energy and National Goals."

15. The first quotation appeared in the caption to a photograph in Iron Eyes Cody as told to Collin Perry, *Iron Eyes: My Life as a Hollywood Indian* (New York: Everest House, 1982); the photograph is in a folio section, which is not paginated. The second quotation is from ABC News and NBC News, April 21, 1978, VTNA.

16. Carter, "Address to the Nation on Energy and National Goals."

17. Marinelli, "Carter's Phony Fuels," 25; Theodore A. Snyder Jr., "Our Energy Future: A Time to Choose," *Sierra*, September/October 1979, 4–5 (quotation on 5).

18. "Carter's Crisis—and Ours," *Progressive*, September 1979, 6–8 (first quotation on 6); Snyder, "Our Energy Future," 4 (second and third quotations); "The Energy Plan," *Newsweek*, July 23, 1979, 27–28 (fourth quotation on 28; fifth quotation on 27). For other leftist critiques of Carter beside this piece in *Progressive*, see also Barry Commoner, *The Politics of Energy* (New

York: Knopf, 1979); and Michael Harrington, *Decade of Decision: The Crisis of the American System* (New York: Simon and Schuster, 1980).

19. "The Energy Plan," 28. See also Marinelli, 25.

20. "Carter's Crisis—and Ours," 7.

21. Christopher Lasch to Patrick Caddell, July 18, 1979, box 20, folder 6, Christopher Lasch papers, Department of Rare Books and Special Collections, University of Rochester Library, reprinted in *Jimmy Carter and the Energy Crisis of the 1970s*, ed. Horowitz, 157–61; all quotations on 160–61.

22. Originally published in the *Philadelphia Inquirer* and syndicated in other newspapers, this cartoon was reprinted in the *Progressive*, October 1979, 11. Other nuclear-themed cartoons by Auth that featured the smiley face were reprinted in the *Progressive*, June 1979, 8; and the *Progressive*, September 1979, 18.

23. Jimmy Carter, "The State of the Union Address Delivered before a Joint Session of the Congress," January 23, 1980, in Woolley and Peters, *The American Presidency Project*, available online at http://www.presidency.ucsb.edu/ws/index.php?pid=33079. On the Carter Doctrine, see Michael T. Klare, *Blood and Oil: The Dangers and Consequences of America's Growing Dependency on Imported Petroleum* (New York: Metropolitan Books, 2004). As Klare explains, Carter chose "to *securitize* oil—that is, to cast its continued availability as a matter of 'national security,' and thus something that can be safeguarded through the use of military force" (12). See also Toby Craig Jones, "America, Oil, and War in the Middle East," *Journal of American History* 99 (June 2012): 208–18.

24. Advertising Council, Alliance to Save Energy Campaign, Story Board titled "Gregory Peck," box 1, folder Forest Fire Prevention; High Blood Pressure Education; National Endowment for the Arts; Carpooling; American Economic System; Energy Conservation, 1978, 13/2/214; Advertising Council, Alliance to Save Energy Campaign, advertisements in file 3084, June 1979, 13/2/207, both in ACA.

25. On Carter's adviser Patrick Caddell's comment about *Network*, see *Jimmy Carter and the Energy Crisis of the 1970s*, ed. Horowitz, 20. On the depiction of "oil sheiks" during the 1970s, see Melani McAlister, *Epic Encounters: Culture, Media, and U.S. Interests in the Middle East since 1945*, updated edition (Berkeley: University of California Press, 2005), chap. 3; and Zaretsky, chap. 2. On anger and the tax revolt, see, for example, Mattson, 176; and Thomas Byrne Edsall with Mary D. Edsall, *Chain Reaction: The Impact of Race, Rights, and Taxes on American Politics* (New York: Norton, 1991), 116–36. Evan McKenzie, *Privatopia: Homeowner Associations and the Rise of Residential Private Government* (New Haven: Yale University Press, 1994), 19.

26. Ronald Reagan, "Remarks Accepting the Presidential Nomination at the Republican National Convention in Dallas, Texas," August 23, 1984, in

Woolley and Peters, *The American Presidency Project*, available online at http://www.presidency.ucsb.edu/ws/?pid=40290. On conservative politicians and frontier imagery, see Robert A. Goldberg, "The Western Hero in Politics: Barry Goldwater, Ronald Reagan, and the Rise of the American Conservative Movement," in *The Political Culture of the New West*, ed. Jeff Roche (Lawrence: University Press of Kansas, 2008), 13–50. On Reagan's rhetoric, see also Rodgers, chap. 1.

27. David Farber, "The Torch Had Fallen," in *America in the Seventies*, ed. Beth Bailey and David Farber (Lawrence: University Press of Kansas, 2004), 9–28, esp. 19–20; and Edward D. Berkowitz, *Something Happened: A Political and Cultural Overview of the Seventies* (New York: Columbia University Press, 2006), 115; AP quotation from 115.

28. Ronald Reagan, "Address Accepting the Presidential Nomination at the Republican National Convention in Detroit," July 17, 1980, in Woolley and Peters, *The American Presidency Project*, available online at http://www.presidency.ucsb.edu/ws/?pid=25970. My argument here builds on Zaretsky, 227–35; and Susan Jeffords, *Hard Bodies: Hollywood Masculinity in the Reagan Era* (New Brunswick, NJ: Rutgers University Press, 1994).

29. Although environmentalists critiqued Carter's energy policies, they offered more positive assessments of other aspects of his environmental record, including his effort to protect wilderness areas and other public lands in Alaska. On these issues, see, for example, James Morton Turner, *The Promise of Wilderness: American Environmental Politics since 1964* (Seattle: University of Washington Press, 2012), 164–76.

CHAPTER TWELVE

1. Robin Toner, "Bush, in Enemy Waters, Says Rival Hindered Cleanup of Boston Harbor," *New York Times*, September 2, 1988; CBS News, September 1, 1988, quoted in Kathleen Hall Jamieson, *Dirty Politics: Deception, Distraction, and Democracy* (New York: Oxford University Press, 1992), 171.

2. "Harbor," Bush-Quayle '88, 1988, directed by Dennis Frankenberry, original air date September 13, 1988. From Museum of the Moving Image, *The Living Room Candidate: Presidential Campaign Commercials, 1952–2008*, http://www.livingroomcandidate.org/commercials/1988/harbor (accessed November 21, 2011).

3. Janet Cawley, "Gentle or Tough? Bush Keeps 'em Guessing," *Chicago Tribune*, October 9, 1988; "The Presidential Debate: Deftly Placed One-Liners Spice Up a Sober Dialogue," *Los Angeles Times*, September 26, 1988.

4. Steven Thomma, "Candidates Faulted on Environment," *Philadelphia Inquirer*, September 22, 1988.

5. Robert V. Percival, "Restoring Regulatory Policy to Serve the Public Interest," in *Winning America: Ideas and Leadership for the 1990s*, ed. Marcus G.

Raskin and Chester W. Hartman (Boston: South End Press, 1988), 49. See also Richard N. L. Andrews, *Managing the Environment, Managing Ourselves: A History of American Environmental Policy* (New Haven: Yale University Press, 1999), esp. 257–58.

6. On neoliberalism, see, among others, David Harvey, *A Brief History of Neoliberalism* (New York: Oxford University Press, 2005); and Lisa Duggan, *The Twilight of Equality? Neoliberalism, Cultural Politics, and the Attack on Democracy* (Boston: Beacon Press, 2003).

7. Ted Steinberg, "Can Capitalism Save the Planet? On the Origins of Green Liberalism," *Radical History Review* 107 (Spring 2010): 7–24 (quotation on 8).

8. On the increasing dominance of market metaphors, see Daniel T. Rodgers, *Age of Fracture* (Cambridge, MA: Harvard University Press, 2011), chap. 2.

9. *Time*, September 22, 1980, cover; James Marsh, quoted in John A. Meyers, "A Letter from the Publisher," *Time*, October 14, 1985, 4.

10. Andrew Szasz, *EcoPopulism: Toxic Waste and the Movement for Environmental Justice* (Minneapolis: University of Minnesota Press, 1994), 43.

11. Gibbs, quoted in Amy M. Hay, "Recipe for Disaster: Motherhood and Citizenship at Love Canal," *Journal of Women's History* 21 (Spring 2009): 117.

12. On toxic waste as a "new mass issue," see Szasz, *EcoPopulism*, 38–68.

13. Szasz, *EcoPopulism*, 129.

14. See Szasz, *EcoPopulism*, 130–33; Meyers, "A Letter," 4.

15. See Emily Brownell, "Negotiating the New Economic Order of Waste," *Environmental History* 16 (April 2011): 262–89. Quotations from Elli Louka, *Overcoming National Barriers to International Waste Trade: A New Perspective on the Transnational Movements of Hazardous and Radioactive Wastes* (Dordrecht, Netherlands: Graham & Trotman / M. Nijhoff, 1994), 112, as quoted in Brownell, 273.

16. *Sex, Lies, and Videotape*, DVD, directed by Steven Soderbergh (1989; Culver City, CA: Sony Pictures Home Entertainment, 1998).

17. NBC News, April 25, 1987, DVD of broadcast from VTNA; CBS News, May 15, 1987, VTNA; *New Yorker*, May 11, 1987, 33.

18. On the *Mobro* as news icon, see W. Lance Bennett and Regina G. Lawrence, "News Icons and the Mainstreaming of Social Change," *Journal of Communication* 45 (September 1995): 20–39; quotation from "A Garbage Scow as Paul Revere," *New York Times*, May 23, 1987.

19. See Steinberg, "Can Capitalism," esp. 16–17; and Ted Steinberg, *Down to Earth: Nature's Role in American History*, 2d ed. (New York: Oxford University Press, 2009), 235–36.

20. See also Catriona Sandilands, "On 'Green' Consumerism: Environmental Privatization and 'Family Values,'" *Canadian Woman Studies* 13 (Spring 1993): 45–47.

21. Patti Jones, "Women Are Waking Up to Their Planet," *Glamour*, May 1990, 270–73 and 323–24; quotations on 271.

22. Allan Mazur and Jinling Lee, "Sounding the Global Alarm: Environmental Issues in the US National News," *Social Studies of Science* 23 (November 1993): 681–720.

23. See Mazur and Lee, esp. 685–93. For a satellite image example, see the photograph that appears in Michael D. Lemonick, "The Heat Is On," *Time*, October 19, 1987, 61. NASA official quoted in Mark Monmonier, *Air Apparent: How Meteorologists Learned to Map, Predict, and Dramatize the Weather* (Chicago: University of Chicago Press, 1999), 211.

24. Mazur and Lee, 691–92; quotation on 691.

25. CBS News, August 10, 1988, VTNA (first quotation); Frank Trippett, "Talking About the Weather," *Time*, August 15, 1988, 20 (second quotation). On the summer of 1988, see also Spencer R. Weart, *The Discovery of Global Warming* (Cambridge, MA: Harvard University Press, 2003), 154–57; and Mazur and Lee, 705–10.

26. *Time*, July 4, 1988, 1 (first quotation); David Brand, "Is the Earth Warming Up?" *Time*, July 4, 1988, 18 (second quotation); *Newsweek*, July 11, 1988; Weart, *The Discovery of Global Warming*, 156 (final quotation).

27. Thomas A. Sancton, "What on Earth Are We Doing?" *Time*, January 2, 1989, 26, 27–28.

28. Robert L. Miller, "From the Publisher," *Time*, January 2, 1989, 3 (first quotation); Don H. Krug, "Teaching Art in the Contexts of Everyday Life," in *Contemporary Issues in Art Education*, ed. Yvonne Gaudelius and Peg Speirs (Upper Saddle River, NJ: Prentice Hall, 2002), 180–97; second quotation on 180.

29. Sancton, "What on Earth," 30.

30. David Brooks, "Journalists and Others for Saving the Planet," *Wall Street Journal*, October 5, 1989.

31. Ad Council, guide and fact sheet, "Questions and Answers about Recycling and Solid Waste Management," 1, in box 2, folder NCP, Recycling Campaign, Sept. 1988, Ad Council, newspaper advertisements, 1961–68, 1987–, 13/2/212, ACA.

32. First quotation from Frederic D. Krupp, form letter to advertising directors, May 1990, in box 3, folder Newspaper Kit, Recycling, May 1990; Ad Council, newspaper advertisements, 1961–68, 1987–, 13/2/212, ACA; other quotations from Environmental Defense Fund Recycling Campaign: Year II Marketing Strategy, November 1989, 5, 16, and 15, in box 2, folder Campaign Review Committee Meeting, Nov. 14, 1989;, Ad Council: Campaign Review Committee File, 1948–97, 13/2/225, ACA.

33. For another example of the media's role in constructing neoliberal models of citizenship, see Laurie Ouellette, "'Take Responsibility for Yourself': Judge Judy and the Neoliberal Citizen," in *Reality TV: Remaking Television Culture*, ed. Susan Murray and Laurie Ouellette (New York: New York University Press, 2004), 231–50.

34. On the emotional history of capitalism, see also Eva Illouz, *Cold Intimacies: The Making of Emotional Capitalism* (Malden, MA: Polity Press, 2007); Joel Pfister, "Getting Personal and Getting Personnel: U.S. Capitalism as a System of Emotional Reproduction," *American Quarterly* 60 (December 2008): 1135–42; and Sarah Banet-Weiser, *Authentic™: The Politics of Ambivalence in a Brand Culture* (New York: New York University Press, 2012).

CHAPTER THIRTEEN

1. Sally Cook and Anne O'Malley, "Forbidden Fruit," *Family Circle*, April 4, 1989, 14–17 (first quotation on 14; third quotation on 16); Brad Darrach, "Meryl," *Life*, December 1987, 72–82; second quotation on 78.
2. "The Greening of Hollywood," *People Weekly*, April 23, 1990, 36.
3. *Time*, September 7, 1981, cover (first quotation); Jeff Rovin, "Thoroughly Modern Meryl," *Ladies' Home Journal*, August 1986, 100, 150–54 (second quotation on 152); Darrach, "Meryl," third quotation on 77; Jack Kroll, "Meryl Streep: Reluctant Superstar," *Ladies' Home Journal*, May 1985, 137, 195–96 (fourth quotation on 195). On Streep's career and public image, see also Karen Hollinger, *The Actress: Hollywood Acting and the Female Star* (New York: Routledge, 2006), 71–99.
4. Bradford H. Sewell, Robin M. Whyatt, et. al., *Intolerable Risk: Pesticides in Our Children's Food* (New York: Natural Resources Defense Council, 1989), quotations on 3 and 37.
5. CBS, *60 Minutes*, "A is for Apple," February 26, 1989, segment included on *NRDC Pesticide Clips*, videotape available from Gelardin New Media Center, Lauinger Library, Georgetown University.
6. Most quotations in this paragraph are from David Fenton, memorandum, excerpted in "How a PR Firm Executed the Alar Scare," *Wall Street Journal*, October 3, 1989. The final quotation is from "The PR Wizard Who Stopped Alar: An Interview with David Fenton," *Propaganda Review*, Summer 1989, 14–17; quotation on 16.
7. Fenton, memorandum, excerpted in "How a PR Firm."
8. First quotation from *Grady Auvil v. CBS "60 Minutes,"* 800 F. Supp. 941 (1992), available online at http://www.cspinet.org/foodspeak/laws/60min2.htm. J. Raloff, "EPA Plans to Ban Carcinogen Daminozide," *Science News*, September 7, 1985, 149. For further background on the Alar controversy, including the role of science in policy debates, see Kerry E. Rodgers, "Multiple Meanings of Alar after the Scare: Implications for Closure," *Science, Technology, & Human Values* 21 (Spring 1996): 177–97; and Sheila Jasanoff, *The Fifth Estate: Science Advisers as Policymakers* (Cambridge, MA: Harvard University Press, 1990), esp. 141–51 and 231–33.
9. Sewell and Whyatt, *Intolerable Risk*, 33 (first, second, and third quotations) and 5 (fourth quotation).

10. Nanci Hellmich and Patrick O'Driscoll, "Streep's New Role," *USA Today*, March 8, 1989. Mothers and Others for Pesticide Limits and Natural Resources Defense Council, television commercials included on *NRDC Pesticide Clips*. These commercials would also provide a phone number and encourage viewers to call to order a copy of *For Our Kids' Sake*, a booklet that provided a more accessible version of the report *Intolerable Risk*. Anne Witte Garland, *For Our Kids' Sake: How to Protect Your Child against Pesticides in Food* (New York: Natural Resources Defense Council, 1989).

11. Conrad Smith, "Responsible Journalism, Environmental Advocacy, and the Great Apple Scare of 1989," *Journal of Environmental Education*, 29 (Summer 1998): 31–37, quotation on 35. CBS News, March 16, 1989, from VTNA; and Bonnie Johnson, "Ms. Streep Goes to Washington to Stop a Bitter Harvest," *People Weekly*, March 20, 1989, 50–51; quotation on 50.

12. John H. Adams and Patricia Adams with George Black, *A Force for Nature: The Story of NRDC and the Fight to Save Our Planet* (San Francisco: Chronicle Books, 2010), 140.

13. CBS News, March 10, 1989, VTNA. For reports of school districts banning apples and apple products, see ABC News, March 14, 1989, VTNA; CBS News, March 14, 1989, VTNA; NBC News, March 14, 1989, VTNA; and Philip Shabecoff, "3 U.S. Agencies, to Allay Public's Fears, Declare Apples Safe," *New York Times*, March 17, 1989.

14. "Even the Apple Has Fallen," *New York Times*, February 6, 1989; Senator Joseph Lieberman, "Apples and Pesticides," *Congressional Record*, 101st Congress, 1st Sess., 135 (March 15, 1989): S2539; and NBC News, March 16, 1989, VTNA.

15. Lizabeth Cohen, *A Consumers' Republic: The Politics of Mass Consumption in Postwar America* (New York: Alfred A. Knopf, 2003), 131 (first quotation); "Alar: Not Gone, Not Forgotten," *Consumer Reports*, May 1989, 290; Rhoda H. Karpatkin, "Memo to Members," *Consumer Reports*, May 1989, 283; and Senator Slade Gorton, "Environmental Irresponsibility," *Congressional Record*, 101st Congress, 1st Sess., 135 (May 4, 1989): S4938.

16. "Fruit Frights," *Wall Street Journal*, March 17, 1989; "Alarmania," *Wall Street Journal*, June 8, 1989; and Elizabeth M. Whelan, "Apple Dangers Are Just So Much Applesauce," *Wall Street Journal*, March 14, 1989. Subsequent examples of conservative writers denouncing the Alar controversy include Warren T. Brookes, "The Wasteful Pursuit of Zero Risk," *Forbes*, April 30, 1990, 161–72; Ronald Bailey, *Eco-Scam: The False Prophets of Ecological Apocalypse* (New York: St. Martin's Press, 1993), 21; and Ben Bolch and Harold Lyons, *Apocalypse Not: Science, Economics, and Environmentalism* (Washington: Cato Institute, 1993), 39–43. For analysis of a previous use of the gendered rhetoric of hysteria to discredit Rachel Carson and *Silent Spring*, see Maril Hazlett, "Voices from the *Spring*: *Silent Spring* and the Ecological Turn in American Health," in *Seeing Nature through Gender*, ed. Virginia J. Scharff (Lawrence: University Press of Kansas, 2003), 103–28, esp. 110–11.

17. Barry Commoner, *Making Peace with the Planet* (New York: Pantheon Books, 1990), 209.

18. Brookes, "The Wasteful Pursuit of Zero Risk," first quotation on 161; Warren Stickle (president of the Chemical Producers and Distributors Association), testimony, US House of Representatives Committee on Agriculture, 1989, as quoted in Rodgers, "Multiple Meanings of Alar," 186.

19. First, second, third, and fifth quotations from Garland, *For Our Kids' Sake*, 30; fourth quotation from Sewell and Whyatt, *Intolerable Risk*, 95.

20. ABC *Nightline*, June 1, 1989, VTNA. On media frames and the tendency to distort, dismiss, and ignore organic agriculture during the 1970s and 1980s, see Warren J. Belasco, *Appetite for Change: How the Counterculture Took on the Food Industry*, 2nd ed. (Ithaca, NY: Cornell University Press, 2007), esp. 167–68. Sinclair quoted in Belasco, 168.

21. Michael Pollan, "Naturally," *New York Times Magazine*, May 13, 2001, 30–37, 57–58, and 63–65 (first quotation on 32; "watershed" on 33); Laura Shapiro, "Suddenly, It's a Panic for Organic," *Newsweek*, March 27, 1989, 24–26.

22. Andrew Szasz, *Shopping Our Way to Safety: How We Changed from Protecting the Environment to Protecting Ourselves* (Minneapolis: University of Minnesota Press, 2007), 2–3.

23. Dana Jackson, "Who Speaks for the Land?" in *Media and the Environment*, ed. Craig L. LaMay and Everette E. Dennis (Washington: Island Press, 1991), 135–46; quotations on 136, 137, and 138.

24. Adams and Adams with Black, *A Force for Nature*, 146.

25. The quotations in the first sentence are from Gregg Mitman, *Reel Nature: America's Romance with Wildlife on Film* (Cambridge, MA: Harvard University Press, 1999), 158 and 178. Other quotations from Earth Island Institute, "The Dolphin Massacre off Our Coast and What You Can Do to Stop It," advertisement, *New York Times*, April 11, 1988.

26. *Lethal Weapon 2*, DVD, directed by Richard Donner (1989; Burbank, CA: Warner Home Video, 1997).

27. Nancy Marx Better, "Green Teens," *New York Times Sunday Magazine*, March 8, 1992, 44, 66–67 (quotations on 66); "Dolphins and Double Hulls," *New York Times*, April 14, 1990; Kenneth R. Clark, "Earth Calling . . . Help!" *Chicago Tribune*, April 22, 1990.

28. Ann Trebbe and Karen Ridgeway, "The Earth and Stars: Environment Is Right for a Chic Cause," *USA Today*, April 12, 1990; Ronald Brownstein, *The Power and the Glitter: The Hollywood-Washington Connection* (New York: Pantheon Books, 1990), 388.

CHAPTER FOURTEEN

1. Tony Dawson, "A Poisoned Dart in Alaska's Heart," *Audubon*, September 1989, 90–91. For background on the oil industry in Alaska and the

Exxon Valdez spill, see Donald Worster, "Alaska: The Underworld Erupts," in his *Under Western Skies: Nature and History in the American West* (New York: Oxford University Press, 1992), 154–224; and Thomas A. Birkland and Regina G. Lawrence, "The *Exxon Valdez* and Alaska in the American Imagination," in *American Disasters*, ed. Steven Biel (New York: New York University Press, 2001), 382–402.

2. Dawson, "A Poisoned Dart," 91. See also Mary Ann Gwinn, "A Deathly Call of the Wild," *Seattle Times*, April 4, 1989.

3. CBS News, NBC News, and ABC News, all from March 30, 1989, DVD available from VTNA; "horrible mistake" from ABC News; "freakish accident" from CBS News.

4. Slavoj Žižek, *Violence: Six Sideways Reflections* (New York: Picador, 2008), 1 and 2.

5. ABC News, March 29, 1989, VTNA.

6. ABC News, March 29, 1989, VTNA.

7. NBC News, April 5, 1989; CBS News, April 6, 1989, both from VTNA.

8. On depictions of individual suffering in other iconic photographs, see also Robert Hariman and John Louis Lucaites, *No Caption Needed: Iconic Photographs, Public Culture, and Liberal Democracy* (Chicago: University of Chicago Press, 2007), esp. 88–91.

9. CBS News, April 11, 1989, VTNA.

10. On media portrayals of rescue centers in the aftermath of the *Exxon Valdez* spill, see also Kathryn Morse, "There Will Be Birds: Images of Oil Disasters in the Nineteenth and Twentieth Centuries," *Journal of American History* 99 (June 2012): 124–34, esp. 133. Quotation from ABC News, April 21, 1989, VTNA. For other examples of TV news coverage of rescue centers, see CBS News, April 2, 1989; CBS News, April 3, 1989; CBS News, April 5, 1989; CBS News, April 7, 1989; NBC News, April 7, 1989; NBC News, April 8, 1989; and CBS News, April 13, 1989, all from VTNA. On rescue centers, see also Ann Larabee, *Decade of Disaster* (Urbana: University of Illinois Press, 2000), esp. 90–97.

11. On the wilderness motif in *Exxon Valdez* coverage, see also Susan Kollin, *Nature's State: Imagining Alaska as the Last Frontier* (Chapel Hill: University of North Carolina Press, 2001), 2–5 and 12–22.

12. Yost quoted on ABC News, CBS News, and NBC News, all on March 30, 1989, all from VTNA; Paul A. Witteman, "First Mess Up, Then Mop Up," *Time*, April 2, 1990, 22; judge quoted on ABC News, CBS News, and NBC News, all on April 5, 1989, all from VTNA; Letterman quoted in Richard Behar, "Joe's Bad Trip," *Time*, July 24, 1989, 43.

13. Exxon, advertisement, *New York Times*, April 3, 1989; Phoebe Wall, letter to editor, *Newsweek*, May 1, 1989, 14.

14. Raymond Loewy, *Industrial Design* (Amsterdam: De Arbeiderspers, 1979), 32 and 198. On Loewy's career, see Glenn Porter, *Raymond Loewy: Designs for a Consumer Culture* (Wilmington, DE: Hagley Museum and Library, 2002).

15. For use of the Exxon logo as part of the background image on network news broadcasts, see, among others, NBC News, April 7, 1989; NBC News, April 10, 1989; and ABC News, April 17, 1989, all from VTNA. For photographs of protesters altering the Exxon logo, see the images in "'They'll Never Get It All,'" *Newsweek*, May 8, 1989, 25; and "'One Way to End a Career,'" *Newsweek*, May 29, 1989, 52. The editorial cartoon by Mike Peters was reprinted in *Newsweek*, September 25, 1989, 13.

16. Citizens for Environmental Responsibility, advertisement in *The Progressive*, August 1989, back cover.

17. First quotation from Robert W. Adler, letter to editor, *Newsweek*, May 1, 1989, 14–15; second quotation from letter read on CBS News, April 11, 1989, VTNA.

18. Citizens for Environmental Responsibility, advertisement in *The Progressive*, August 1989, back cover; Thomas L. Friedman, *The Lexus and the Olive Tree*, rev. ed. (New York: Farrar Straus Giroux, 2000), 161.

19. Lawrence Rawl quoted in Richard Behar, "Exxon Strikes Back," *Time*, March 26, 1990, 62; Frank Iarossi (other Exxon official) quoted in Jerry Adler, "Alaska after Exxon," *Newsweek*, September 18, 1989, 54.

20. *U.S. News & World Report*, September 18, 1989, cover; Michael Satchell and Betsy Carpenter, "A Disaster That Wasn't," *U.S. News & World Report*, September 18, 1989, first and last quotations on 69 and 62. The phrases "special absorbent pads" and "one by one" are from the caption to a photograph in Adler, "Alaska after Exxon," 54.

21. *National Geographic*, January 1990, cover; Bryan Hodgson, "Alaska's Big Spill: Can the Wilderness Heal?" *National Geographic*, January 1990, 42.

22. On these issues in relation to the *Deepwater Horizon* spill, see Rob Nixon, *Slow Violence and the Environmentalism of the Poor* (Cambridge, MA: Harvard University Press, 2011), 21–22, 273–74, and 276.

23. On the Oil Pollution Act of 1990, see Thomas A. Birkland and Regina G. Lawrence, "The Social and Political Meaning of the *Exxon Valdez* Oil Spill," *Spill Science & Technology Bulletin* 7 (June 2002): 17–22. On the limits of this legislation and the continuing failure to regulate the oil industry, see Stephen Haycox, "'Fetched Up': Unlearned Lessons from the *Exxon Valdez*," *Journal of American History* 99 (June 2012): 219–28, esp. 223 and 228.

24. Murray Bookchin, "Death of a Small Planet," *The Progressive*, August 1989, 19–23; quotations on 19 and 19–20.

25. Bill McKibben, "The Exxon Valdez as a Metaphor," *New York Times*, April 7, 1989.

26. McKibben, "The Exxon Valdez."

27. Most quotations in this paragraph are from *Time*, December 25, 1989, 2. The Berlin Wall quotation appears on 42. The Dawson photograph appears on 56–57.

CHAPTER FIFTEEN

1. *The Earth Day Special*, VHS, directed by Dwight Hemion (1990; Burbank, CA: Warner Home Video, 1990).

2. Quotation from Jonathan Storm, "As Earth Day Grows Near, Ecology Takes Over the Tube," *Philadelphia Inquirer*, April 16, 1990. For another scholarly analysis of *The Earth Day Special*, see Michael X. Delli Carpini and Bruce A. Williams, "'Fictional' and 'Non-Fictional' Television Celebrates Earth Day; or, Politics Is Comedy plus Pretense," *Cultural Studies* 8 (January 1994): 74–98, esp. 88–95.

3. Connie Koenenn, "A Hard Sell to Rescue Planet Earth," *Los Angeles Times*, October 26, 1989.

4. On Hayes, see, for example, James L. Franklin, "Earth Day Message: Get Tough," *Boston Globe*, April 15, 1990; Soroka quoted in Amy Wallace, "Earth Day Is Back to Face Growing Challenges, Dangers," *Los Angeles Times*, January 29, 1990.

5. On Earth Day 1990 and the discourse of individual responsibility, see also Robert Gottlieb, *Forcing the Spring: The Transformation of the American Environmental Movement* (Washington: Island Press, 1993), 201–4; Timothy W. Luke, *Ecocritique: Contesting the Politics of Nature, Economy, and Culture* (Minneapolis: University of Minnesota Press, 1997), chap. 6; Michael F. Maniates, "Individualization: Plant a Tree, Buy a Bike, Save the World?" *Global Environmental Politics* 1 (August 2001): 31–52; and Adam Rome, *The Genius of Earth Day: How a 1970 Teach-In Unexpectedly Made the First Green Generation* (New York: Hill and Wang, 2013), 276–80. On neoliberal models of citizenship, see also Laurie Ouellette, "'Take Responsibility for Yourself': Judge Judy and the Neoliberal Citizen," in *Reality TV: Remaking Television Culture*, ed. Susan Murray and Laurie Ouellette (New York: New York University Press, 2004), 231–50.

6. On neoliberalism and the emotional life of capitalism, see Sarah Banet-Weiser, *Authentic™: The Politics of Ambivalence in a Brand Culture* (New York: New York University Press, 2012). On environmental justice as a challenge to neoliberal models of citizenship, see Julie Sze, *Noxious New York: The Racial Politics of Urban Health and Environmental Justice* (Cambridge, MA: MIT Press, 2007), 9–11.

7. ABC News, January 24, 1990, VTNA.

8. ABC News, January 23, 1990, VTNA.

9. Hayes quoted in David Graham, "Environmentally Safe Products to Get 'Green Seal' of Approval," *San Diego Union*, June 16, 1990.

10. Bernstein quoted in Diane Haithman, "TV Gets Down to Earth," *Los Angeles Times*, April 15, 1990.

11. On tree-planting children, see also Holly Stocking and Jennifer Pease Leonard, "Fingering the Bad Guys: Ironies in Environmental Journalism," *Newspaper Research Journal* 11 (Fall 1990): 2–11, esp. 3.

12. My argument here draws on Lauren Berlant, *The Queen of America Goes to Washington City: Essays on Sex and Citizenship* (Durham, NC: Duke University Press, 1997), esp. 1–53. In chapter 1, I engaged with Berlant's theory of infantile citizenship to argue that SANE's images of vulnerable children revealed the dangers to permeable ecological bodies and thereby made new demands upon the state to protect the citizenry from harm. In this chapter, my argument more closely corresponds to Berlant's analysis, as I am suggesting that the imagery of children in *The Earth Day Special* and other popular texts deflected attention from power relations.

13. Stacy Alaimo, *Undomesticated Ground: Recasting Nature as Feminist Space* (Ithaca, NY: Cornell University Press, 2000), 174.

14. CBS News, April 21, 1990, VTNA. On affective empowerment, see also Lawrence Grossberg, *We Gotta Get Out of This Place: Popular Conservatism and Postmodern Culture* (New York: Routledge, 1992), 86.

15. T. J. Jackson Lears, "From Salvation to Self-Realization: Advertising and the Therapeutic Roots of the Consumer Culture, 1880–1930," in *The Culture of Consumption: Critical Essays in American History, 1880–1980*, ed. Richard Wightman Fox and T. J. Jackson Lears (New York: Pantheon, 1983), 3–38; quotations on 27. Christopher Lasch, *The Culture of Narcissism: American Life in an Age of Diminishing Expectations* (New York: W. W. Norton, 1979).

16. Heather Rogers, *Gone Tomorrow: The Hidden Life of Garbage* (New York: The New Press, 2005), 174 and 176.

17. On recycling programs as corporate subsidy, see Bartow J. Elmore, "The American Beverage Industry and the Development of Curbside Recycling Programs, 1950–2000," *Business History Review* 86 (Autumn 2012): 477–501.

18. Michael Lerner, "Critical Support for Earth Day: An Editorial," *Tikkun*, March/April 1990, 48; Commoner quoted in Dena Kleiman, "How Do You Fix a Broken Planet," *New York Times*, April 25, 1990.

19. Chavis quoted in Giovanna Di Chiro, "Nature as Community: The Convergence of Environment and Social Justice," in *Uncommon Ground: Rethinking the Human Place in Nature*, ed. William Cronon (New York: Norton, 1995), 298–320; quotation on 304.

20. Quotations from Robert Cahn and Patricia Cahn, "Did Earth Day Change the World?" *Environment*, September 1990, 16–20, 36–43; quotations on 36. On the toxic tour, see also Gary Cohen, "It's Too Easy Being Green," *Social Policy* 21 (Summer 1990): 28–29.

21. United Church of Christ Commission for Racial Justice, *Toxic Wastes and Race in the United States: A National Report on the Racial and Socio-economic Characteristics of Communities with Hazardous Waste Sites* (New York: United Church of Christ, 1987).

22. The "new rainbow of hues" quotation is from Clarence Page, "Minorities 'Going Green' as Ecology Gets Rainbow Voice," *Chicago Tribune*, April 18, 1990.

23. Shepard quoted in John Leland, "Concert Rocked and Preached," *Newsday* (Long Island, NY), April 23, 1990; ABC News, CBS News, and NBC News, April 22, 1990, all from VTNA. On Shepard and West Harlem Environmental Action (WE ACT), see Sze, *Noxious New York*. As Sze notes on 99–100, WE ACT also used the iconic gas mask in its environmental justice campaigns. Rather than signifying universal vulnerability, though, the imagery focused on the unequal levels of environmental risk in primarily African American and Latino neighborhoods.

24. Mednick quoted in Bruce Horovitz, "Designers Pitch in to Help Out Mother Earth," *Los Angeles Times*, April 2, 1990. For analysis of how the discourse of global environmentalism obscures global power inequities, see Giovanna Di Chiro, "Beyond Ecoliberal 'Common Futures': Environmental Justice, Toxic Touring, and a Transcommunal Politics of Place," in *Race, Nature, and the Politics of Difference*, ed. Donald S. Moore, Jake Kosek, and Anand Pandian (Durham, NC: Duke University Press, 2003), 204–32; and Sheila Jasanoff, "Heaven and Earth: The Politics of Environmental Images," in *Earthly Politics: Local and Global in Environmental Governance*, ed. Sheila Jasanoff and Marybeth Long Martello (Cambridge, MA: MIT Press, 2004), 31–52. On the end of the Cold War and the discourse of global environmentalism, see also Frank Uekoetter, "The End of the Cold War: A Turning Point in Environmental History?" in *Environmental Histories of the Cold War*, ed. J. R. McNeill and Corinna R. Unger (Washington: German Historical Institute; New York: Cambridge University Press, 2010), 343–51. For one example of post–Cold War environmental exuberance, see Hugh Sidey, "The Issue That Won't Wash Away," *Time*, March 26, 1990, 21.

25. Garry Wills, "Earth Day Reminds Us to Save Planet," *San Diego Tribune*, April 18, 1990; "The Blackest Town In the World," *Time*, March 19, 1990, 34; CBS News, April 17, 1990, VTNA. Wills was referring to photographs that appeared in Anastasia Toufexis, "Legacy of a Disaster," *Time*, April 9, 1990, 68–70.

26. "Eastern Europe: The Polluted Lands," photo essay by Antonin Kratochvil with text by Marlise Simons, *New York Times Magazine*, April 29, 1990, 30–35.

27. Francis Fukuyama, "The End of History?" *National Interest*, Summer 1989, 3–18; and Fukuyama, *The End of History and the Last Man* (New York: The Free Press, 1992). On the dominance of market metaphors, see Daniel T. Rodgers, *Age of Fracture* (Cambridge, MA: Harvard University Press, 2011), chap. 2.

28. Emily Brownell, "Negotiating the New Economic Order of Waste," *Environmental History* 16 (April 2011): 262–89 (quotations on 276 and 281).

29. Ashley Dawson, "Slow Violence and the Environmentalism of the Poor: An Interview with Rob Nixon," *Social Text* blog, August 31, 2011, http://www.socialtextjournal.org/blog/2011/08/slow-violence-and-the-environmentalism-of-the-poor-an-interview-with-rob-nixon.php.

30. Ken Conca and Geoffrey D. Dabelko, "Twenty-Five Years of Global Environmental Politics," in *Green Planet Blues: Environmental Politics from Stockholm to Kyoto*, ed. Ken Conca and Geoffrey D. Dabelko, 2nd ed. (Boulder, CO: Westview Press, 1998), 6 (first quotation). The "not up for negotiation" is sometimes attributed to Bush, but originated in a *Time* article as a summary of the position of US delegates. See Philip Elmer-DeWitt, "Summit to Save the Earth: Rich vs. Poor," *Time*, June 1, 1992, 42–58; second quotation on 58.

31. Mark Hertsgaard, "Covering the World; Ignoring the Earth," *Rolling Stone*, November 16, 1989, reprinted in *The Rolling Stone Environmental Reader* (Washington: Island Press, 1992), 3–14; quotation on 11.

32. For critiques of the Clean Air Act of 1990, see Ted Steinberg, *Down to Earth: Nature's Role in American History*, 2nd edition (New York: Oxford University Press, 2009), 254–55, and Cohen, "It's Too Easy Being Green," 25–26.

33. On the failure of the Clean Air Act to regulate greenhouse gas emissions and the decline in fuel economy during the 1990s, see Tom McCarthy, *Auto Mania: Cars, Consumers, and the Environment* (New Haven: Yale University Press, 2007), 244–45. Michael T. Klare, *Blood and Oil: The Dangers and Consequences of America's Growing Dependency on Imported Petroleum* (New York: Metropolitan Books, 2004), 50.

34. On SUV sales during the 1990s, see McCarthy, *Auto Mania*, chap. 12.

35. On climate change denial, see Naomi Oreskes and Erik M. Conway, *Merchants of Doubt: How a Handful of Scientists Obscured the Truth on Issues from Tobacco Smoke to Global Warming* (New York: Bloomsbury Press, 2010), chap. 6. For critiques of the class-based elitism of green consumerism, see also Banet-Weiser, *Authentic™*, chap. 4; Josée Johnston, "The Citizen-Consumer Hybrid: Ideological Tensions and the Case of Whole Foods Market," *Theory and Society* 37 (June 2008): 229–70; and Christopher C. Sellers, *Crabgrass Crucible: Suburban Nature and the Rise of Environmentalism in Twentieth-Century America* (Chapel Hill: University of North Carolina Press, 2012), 288 and 294–95.

CONCLUSION

1. Jay Leno, *Tonight Show*, as quoted in Daniel Kurtzman, "Global Warming Jokes: Late-Night Jokes about Global Warming," About.com Political Humor, http://politicalhumor.about.com/od/environment/a/globalwarming .htm; Richard Cohen, "A Campaign Gore Can't Lose," *Washington Post*, April 18, 2006; Pat Aufderheide, review of *An Inconvenient Truth*, *Cineaste*, Winter 2006, 50–52 (quotation on 50).

2. William Booth, "Al Gore, Rock Star," *Washington Post*, February 25, 2007 (first quotation). The Republican strategist (Frank Luntz) is quoted in Tim Dickinson, "Run, Al, Run," *Rolling Stone*, February 8, 2007, 42–45;

quotation on 43. On the Nobel Peace Prize, see "Gore Shares Peace Prize for Climate Change Work," *New York Times*, October 13, 2007. I borrow the phrase "carbon warrior" from an article by the political scientist Kate Ervine; see full citation in note 17.

3. *Time*, April 3, 2006, cover; *An Inconvenient Truth*, DVD, directed by Davis Guggenheim (2006; Hollywood: Paramount Pictures, 2006); *Vanity Fair*, May 2007, cover.

4. Jeffrey Kluger, "The Tipping Point," *Time*, April 3, 2006, caption on 29. For further commentary on the polar bear as an icon of global warming, emphasizing its status as charismatic megafauna, see Denis Cosgrove, "Images and Imagination in 20th-Century Environmentalism: From the Sierras to the Poles," *Environment and Planning A* 40 (2008): 1862–80, esp. 1877–78.

5. Environmental Defense Fund and the Ad Council, "Train," (2006). The PSA can be viewed at http://www.youtube.com/watch?v=s-_LBXWMCAM; accessed on March 6, 2012.

6. "Commentary with Director," *An Inconvenient Truth*; David Denby, "Tuning In," *The New Yorker*, June 12, 2006, 160–63; quotation on 163.

7. Denby, "Tuning In," 163; Aufderheide, review of *An Inconvenient Truth*, 51.

8. Roger Ebert, "An Inconvenient Truth," *Chicago Sun-Times*, June 2, 2006. On intergenerational ethics and global warming, see also Stephen M. Gardiner, "A Perfect Moral Storm: Climate Change, Intergenerational Ethics and the Problem of Moral Corruption," *Environmental Values* 15 (August 2006): 397–413.

9. A. O. Scott, "Warning of Calamities and Hoping for a Change in 'An Inconvenient Truth,'" *New York Times*, May 24, 2006.

10. "Commentary with Producers Laurie David, Lawrence Bender, Scott Z. Burns, and Leslie Chilcott," *An Inconvenient Truth*.

11. Rob Nixon, *Slow Violence and the Environmentalism of the Poor* (Cambridge, MA: Harvard University Press, 2011).

12. For thoughtful critiques of whole earth imagery, see Giovanna Di Chiro, "Beyond Ecoliberal 'Common Futures': Environmental Justice, Toxic Touring, and a Transcommunal Politics of Place," in *Race, Nature, and the Politics of Difference*, ed. Donald S. Moore, Jake Kosek, and Anand Pandian (Durham, NC: Duke University Press, 2003), 204–32; and Sheila Jasanoff, "Heaven and Earth: The Politics of Environmental Images," in *Earthly Politics: Local and Global in Environmental Governance*, ed. Sheila Jasanoff and Marybeth Long Martello (Cambridge, MA: MIT Press, 2004), 31–52.

13. Yates McKee, "Art and the Ends of Environmentalism: From Biosphere to the Right to Survival," in *Nongovernmental Politics*, ed. Michel Feher with Gaëlle Krikorian and Yates McKee (New York: Zone Books, 2007), 539–83; quotation on 558–59. See also Noël Sturgeon, "Penguin Family Values: The Nature of Planetary Environmental Reproductive Justice," in *Queer Ecologies: Sex, Nature, Politics, Desire*, ed. Catriona Mortimer-Sandilands

and Bruce Erickson (Bloomington: Indiana University Press, 2010), 102–33, esp. 119–22. On environmental and indigenous activism in the Arctic, see also *Arctic Voices: Resistance at the Tipping Point*, ed. Subhankar Banerjee (New York: Seven Stories Press, 2012).

14. See Alexa Weik von Mossner, "Reframing Katrina: The Color of Disaster in Spike Lee's *When the Levees Broke*," *Environmental Communication* 5 (June 2011): 146–65; Robert D. Bullard and Beverly Wright, preface to *Race, Place, and Environmental Justice after Hurricane Katrina: Struggles to Reclaim, Rebuild, and Revitalize New Orleans and the Gulf Coast* (Boulder, CO: Westview Press, 2009), ed. Robert D. Bullard and Beverly Wright, xix–xxii.

15. Michael Pollan, "Why Bother?" *Saturday Evening Post*, September/October 2008, 42–45; quotation on 42. See also John M. Meyer, "*Another* Inconvenient Truth," *Dissent* (Fall 2006): 95–96.

16. All quotes from ABC 13 News, KTRK-TV, Houston, "Al Gore's 'Inconvenient Truth'?—$30,000 Utility Bill," http://abclocal.go.com/ktrk/story?section=news/national_world&id=5072659, accessed on May 29, 2012.

17. For two compelling critiques of carbon offsetting, see Heather Rogers, *Green Gone Wrong: How Our Economy Is Undermining the Environmental Revolution* (New York: Scribner, 2010), 149–77; and Kate Ervine, "The Politics and Practice of Carbon Offsetting: Silencing Dissent," *New Political Science* 34 (March 2012): 1–20. On Gore's long-standing embrace of neoliberal values and market-oriented environmentalism, see also Andrew Ross, "Earth to Gore, Earth to Gore," *Social Text* 41 (Winter 1994): 1–10.

18. Rogers, *Green Gone Wrong*, 152.

19. Ervine, "The Politics and Practice of Carbon Offsetting," 1, 7, and 19.

20. I viewed the Philips commercial at http://www.youtube.com/watch?v=F1WhVni081U, where it is simply titled "Philips Energy Efficient Lighting Commercial," (2006), accessed on May 13, 2012; DDB is credited as the advertising agency that produced the commercial in Joan Voight, "Why 'Less is More' is Fit for the Zeitgeist," *Adweek*, February 26, 2007, http://www.adweek.com/news/advertsing/why-less-more-fit-zeitgeist-88071, accessed on May 13, 2012. I first learned of the commercial from the analysis of it in Sturgeon, "Penguin Family Values," 127.

21. The quotations are from Jeremy Osborn, operations director of 350.org, as quoted in Elana Schor, "Advocacy: The Education of a Climate Upstart with a 'Weird' Name," *E & E Daily*, December 13, 2013, available online at http://eenews.net/stories/1059991802. See also Bill McKibben, "An Inconvenient Solution," *The Nation*, November 19, 2009, available online at http://www.thenation.com/article/inconvenient-solution; and Bill McKibben, "A Moral Atmosphere," *Orion*, March/April 2013, available online at http://orionmagazine.org/index.php/articles/article/7378.

22. McKibben quoted in DJ Spooky, "Performing Climate Change: DJ Spooky Talks with Bill McKibben," May 20, 2013, available online at http://

creativetimereports.org/2013/05/20/dj-spooky-and-bill-mckibben-350 -org/. Jamie Henn, communications director of 350.org, discusses the use of Flickr (including the figure of twenty thousand uploaded photos) in "Jamie Henn Takes Us Inside 350.org," podcast on Jon Christensen blog, http://christensenlab.net/jamie-henn-takes-us-inside-350-org/.

23. "350.org: Because the World Needs to Know," uploaded to YouTube by 350.org on June 9, 2008, http://www.youtube.com/watch?v=s5kg1oOq9tY.

24. See Bill McKibben, "The Fossil Fuel Resistance," *Rolling Stone*, April 11, 2013, available online at http://www.rollingstone.com/politics/news/the -fossil-fuel-resistance-20130411.

25. On Tar Sands Action's use of new media, see Robert Wilson, "The Necessity of Activism," *Solutions* 3 (July 2012): 75–79, available online at http:// www.thesolutionsjournal.com/node/1129. First quotation from CBC News, "Naomi Klein Arrested at D.C. Pipeline Protest," September 2, 2011, http://www.cbc.ca/news/world/story/2011/09/02/world-naomi-klein -arrested-at-white-house-keystone-pipeline-protest.html. Second quotation from Tar Sands Action, "Big News Day: Tar Sands Action Endorsed by Al Gore, Tops Google News," September 1, 2011, http://www.tarsandsaction .org/big-news-day-tar-sands-action-endorsed-al-gore-tops-google-news/.

Index

Italicized page numbers refer to illustrations.